ECLECTIC PRAGMATISM

Is There a God?

By

Charles H. Peterson

ISBN: 1-4033-5969-5 (e-book)
ISBN: 1-4033-5970-9 (Paperback)
ISBN: 1-4033-5971-7 (Hardcover)

Library of Congress Control Number: 2002093614

This book is printed on acid free paper.

Printed in the United States of America
Bloomington, IN

1stBooks - rev. 11/14/03

ACKNOWLEDGMENTS

To all who have contributed
directly and indirectly
knowingly and unknowingly
to my knowledge and understanding
of the options we have
in answering the fundamental question
we privately face. . .

CONTENTS

Chapter 1. Introduction

<u>External Pressures</u>

With a world population of five billion and a crude birth rate of 26 per 1000, every year a fresh crop of 130,000,000 babies comes into this world, ignorant of the diverse skills, practices, customs, laws and beliefs of the societies into which they have been born. Crude birth rate means the number of births per year per 1000 persons without accounting for how many of the 1000 persons are female and in the child-bearing years. Consider:

- The population of the continental United States in 1940 was about 130,000,000.

- The baby crop is increasing by more than 2 million every year, since the crude death rate of about 9 per 1000 makes the net population increase about 17 per 1000, or 85 million.

What becomes of all these babies? Many die young of birth defects, disease, malnutrition, or violence. The survivors become slaves, laborers, merchants, mothers, professionals, bureaucrats, or rulers. Some become incapacitated, invalided, or insane. Others become thieves, smugglers, prostitutes, rapists, or killers. The rest struggle to earn a "legitimate" living with their respective talents, training, and tools.

Whatever the year of our birth, each of us has to learn many things simply to survive. For several months, we cannot use our own arms and legs to move our bodies from one place to another. Our language is mainly various cries and wails of displeasure and unmet needs: I'm hungry, uncomfortable, hurting, etc. Learning to communicate is thus a slow process until we master a spoken language.

Ah, but we have a lot of help. From birth, parents, siblings, relatives, friends, and even strangers tell us what to do and when and how to do it:

It's time to eat, bathe, sleep.
Wash your hands, brush your teeth, comb your hair.
Do your chores, mind your manners, go to school.
Obey your parents, your teachers, the law.
Get a job, get married, have children.
Go to war and kill the enemy, i.e., people who are different.

They also urge us, expect us, and compel us to live by certain ideas:

> Believe in God, Allah, Brahma, Buddha, Tao, Manitou, etc.
> Live righteously.
> Kill/love your enemies.

But we are not told why nor do we understand why. And all this help does divide us into differing, competing, even hostile groups.

Some of this guidance seems reasonable. As children, we have a tendency, perhaps a drive, to copy what others do. We tend to submit and even accept the directions and commands of our parents, our teachers, and other adults. We do this because we are striving to survive in response to some inner drive or command and because sometimes the guidance is candy-coated with warmth, affection and love. Sometimes we comply because of fear: of rejection, of punishment, of the unknown. And sometimes because we haven't realized that we can say no.

Some of it, particularly in the area of religion,

a) Is incomprehensible;

An omnipotent God created the Universe, but we can only say God was not created, did not create Himself, but existed forever.

b) Is confusing;

- God/Jesus relationship

    John 10:30 "I (Jesus) and my Father are one."

    John 5:30 "I (Jesus) can of mine own self do nothing. . . ."

    (References such as these are from **KJV**.)

c) And is conflicting.

- Human planning

    Matthew 6:34 "Take. . .no thought for the morrow. . . ."

    Luke 14:28 In war or in erecting a building we are to plan ahead.

- God's plan

Isaiah 14:24 ". . .as I (God) have thought, so it shall come to pass; and as I have purposed, so shall it stand. . . ."

Ecclesiastes 9:11 "I returned, and saw. . .that. . .time and chance happeneth to them all (men)."

Question: How can we reconcile these to give us understanding and guidance?

Science is not exempt. To a faithful devotee of the religion of Science the following must be heresy:

Einstein "I shall never believe that God plays dice with the world (**FQ**, p 950b).

Einstein "Religion without science is bland. Science without religion is lame." (**GNP**, Quotation inside front cover.)

We may still comply because somehow we feel the need to be accepted by a group, to have comfort from our friends, and to have their approval. But it all amounts to a set of external pressures on us to conform to the ideas, beliefs, expectations and needs of others, both living and dead. We are subjected to a tidal wave of information from the television, movies, radio, newspapers, magazines and books, much of it aimed at getting us to buy things or "buy into" ideas. We are assailed by so many ideas that our emerging sense of self is threatened with submergence, as a person drowning in the sea. Some of us resist, rebel, and run to ways to avoid complete surrender.

Individuality

And so we reach an important milestone in our lives: an awareness of ourselves as a separate person, with our own thoughts, desires, capabilities, and limitations. We go to a lot of trouble striving to make independent lives for ourselves. There is in all of us something intangible that is forever saying "I", "I", "I" - sometimes even screaming it. We don't seem to question it, although we strongly resist other intangible concepts in our lives.

The Western world urges us to develop our individuality. We are told we are different, special, unique. We are told that in the Christian religion there is a God with a unique plan for each of us. These recurrent notions of God and the question of the place of religion in our lives are particularly disturbing.

3

Yet we can see there are many ways in which we are all alike. We resemble each other more than we do an ape, an asp, or an ant. We share a common pattern of birth - growth - decline - death. Most add another event: reproduction. But between our two end points in time, all we really have is the opportunity to have a set of experiences. We come into the world with nothing; i.e., with none of the material things of this world except a body. We take none of those material things with us, and leave behind maybe a bigger body.

We will not have all the experiences we may dream of because we are born into different circumstances, we don't have enough time, and others are competing for the same experiences. So there are inevitable conflicts. At any one time there can be only one U. S. president, so 259,999,999 people in America are denied the rewards of reaching this goal. Similarly, there can be only one winner of the horse race, one winner of the beauty contest, one winner of the business contract. So we must accept at least such limitations on our desire for individualism. We are also forced to endure unwanted experiences.

Mental Pictures

How do we respond to these pressures? Initially, we are aware mostly of the physical world. We learn about hunger, scraped knees, hot stoves, freezing winters, sunburn. We learn about birth, sickness, and death. We learn about frustration, anger, hate, fear, kindness, and love. As we learn, we build a picture in our minds of what the world is like, how to deal with it and the people in it.

Our picture is formed from our:

- experiences, which we accept as true;
- selections from what we've been told; and
- interpretations of these inputs.

Insinuated into this picture are images and ideas picked up from the outside world of which we often are not even aware. This is not surprising. No one of us has a mental gatekeeper that is 100% conscious at all times, evaluating the information that we take in.

Now there are some mysteries.

- Something within us drives us to make a consistent coherent picture. "We" crave certainty. "We" abhor confusion and loose ends.

4

- Something within us selects information. "We" reject the pieces that in our judgment don't "fit", even though they are quite real and true to others. "We" distort and deny the evidence of our senses. "We" even invent information so that "we" might have a picture that shows us as good and right.

- Something within us organizes the information. Our brains record the selected information in a retrievable manner. (This alone is a major mystery, if not a miracle.) Our brains associate it with related information. If it seems "reasonable" (fits other ideas we have accepted), our minds retain it.

We have both short term and long term memories. The former is like the working memory in a computer, which is an array of on-off switches that code information. This information is erased when the computer is turned off. Many things need to be remembered for only a short time and should be forgotten after that time. But other things need to be remembered longer.

Most of the time we are not conscious of these decisions. It is as if we have a mental electrician who somehow decides what information to retain, where to put it, and makes appropriate connections in our brains. But just who is or what is our electrician? And how does he decide what to keep, how to organize it, and how to record it?

Question: Is memory stored in material brains in the form of various chemical compounds or in electromagnetic fields generated by the arrangement of the atoms in those compounds?

Perhaps everything we receive goes first into a short term memory and later gets transferred into a long term memory. Regardless of the form in which an image is received into our short term memory, somehow it must get replicated into our long term memory either as a duplicate set of chemical compounds or as a duplicate electromagnetic field. Perhaps if there is no similar memory in long term memory with which the image can be associated within a short period of time, it is attenuated and lost.

Something within us issues instructions to make the transfer. In what form is the transfer made? If chemical compounds are involved, one can imagine the instructions are to transfer those compounds atom by atom. How are they tagged so as to be reassembled in the correct order? Or perhaps there is short term and long term memory capability distributed all over the brain so no physical transfer is needed.

5

This could operate like computer instructions. On executing a given command, control simply passes automatically to the next command. But even this step has to be in some master program. The question is really: How did such a set of instructions ever arise? How and where are they "remembered"?

Our picture is always incomplete because every day we get more information. It is defective because our perceptions are incomplete and inaccurate and our interpretations contain many misconceptions and errors. Despite these objections, we need to construct such pictures because they stabilize our lives, make the world more predictable, and help us handle day-to-day situations, as well as guess about the future.

What is the picture we are urged to build?

- There is a worldly part with ideas like these:

  If we spend many years in school and learn an academic set of skills, we will get a good job that pays for all the material things we want. By hard work we will reach the top of our field.

  Anyone of us can be president. Everyone can marry a beautiful loving wife or a handsome loving husband and live happily ever after with a family of x ideal children, a dog, a house, and all kinds of other material things.

- There is a conceptual part.

  Believe in our God that our troubles may be taken care of. (Christianity) (Islam) (Judaism) (Hinduism) (Buddhism) (etc.) is the only true path to the good life here on Earth. When we die, we will go to Paradise and live in eternal bliss.

The worldly part is the easier to formulate, to understand, and to work with. The conceptual part involves continuing conflicts, contradictions, and compromises as we struggle to reconcile what we experience with what we are asked to believe.

Actual Pictures

How do these pictures fit with what we actually experience? Without going into detail, we can list some of the major trends we can see behind the headlines. Let us look first at trends we might call positive. We see increased productivity, and greater control over at least some aspects of our environment.

- Technological - Accelerating rate of advances in knowledge

  Food production
  - o Insect and disease resistant food plants
  - o Increased production per plant and per acre

  Human and plant biology
  - o Increased understanding of biochemistry
  - o Genetic engineering, cloning
  - o More effective medicines and techniques

  Ever faster and more powerful computers

  Miniaturization of devices

  Communications
  - o Instantaneous worldwide transmission of ideas
  - o Optical fibers instead of copper wires

- Personal

  More leisure time and more ways of spending (enjoying?) it, at least for some

  More freedom (for some)

- Social

  Continued, yet slow, progress toward achieving universal human rights

These seem to confirm that much progress is being made in improving our control over our material environment and the quality of our material lives. Now let's look at some negative trends. We see forced change, cruelty, injustice, and escalating violence.

- Individual

  Decreased economic security
  - o Intensified competition for jobs

7

- Elimination of thousands of jobs, destroying the security, frustrating the dreams, and undermining the beliefs of those workers.

Sense of emptiness and meaninglessness of life
- Diminishing hopes of more than just us ordinary humans
- Anonymity and loneliness
- Dropouts from school, work, society
- Withdrawal into a fantasy world - movies and TV, drugs and alcohol
- Suicides

Intensified craving for stimulation
- Indulgence even at risk of life - parachuting, hang gliding, bungee jumping
- Lust for instant gratification of our pleasure drive - eating, drinking, sex
- Lust for gratification from violence
  Contact sports, prize fights, bull fights
  Hunting for the sake of killing
  Vandalism, robbery, murder

Increasing incidence of individual violence
- Homicides
- Violent gangs

Erosion of freedoms won by the sufferings and deaths of millions of people like ourselves over the centuries

- Family

Shattering of the historical family structure
- Decline in authority of fathers/males
- Increasing child abuse, even to murder
- Divorce rate at 50% of all new marriages
- Increasing number of single parents

Lifestyles
- Solitary
- Deviant sexualities
- Normless and multinormic society

Declining attendance at traditional churches
- Rejection of traditional messages

- o Competing faiths, including secularism
- o Cults, including Satanism
- o Reappearance of paganism with animal and human sacrifice

- Society

Increasing polarization of society

The opulent	The destitute
The employed	The unemployable
The educated	The uneducable
The housed	The homeless

Increasing millions not in "mainstream" America
Over age 65	25
Unemployed	7
Disabled workers	3
Imprisoned	1
Institutionalized (mental, health, elder care)	1?

Increasing conflicts: humans are killing humans.
- o Economic oppression/Slavery
- o Political oppression
- o Religious conflicts
  - Traditional Christianity and Science
  - Sectarian conflicts

Increasing medical problems
- o Cardio-vascular disorders, cancer, arthritis, diabetes, Alzheimer's disease
- o Hepatitis, influenza, AIDS and other viruses
- o Drug-resistant bacteria: tuberculosis, staphylococci, E. coli

- Environment

Depletion of all natural resources
- o Energy sources: wood, coal, oil, gas, uranium
- o Minerals
- o Arable land, topsoil, rain forests
- o Edible fish

Increasing production of garbage

Pollution of air, water, land and ocean
- o Global warming
- o Destruction of the ozone layer
- o Upset of balances in ecosystems

Emergence of aggressive species of insects

Natural catastrophes: earthquakes, tornados, floods, and droughts

Question: How do we explain all these events and trends? Are we helpless to affect them? Our secular leaders seem concerned only politically. Where are the voices of our spiritual leaders?

Personal Impact

So we see "good" things happening and "progress" being made but we also see continuation, if not escalation, of violent conflict. In the past, many of us have been able to comfort ourselves with the thought that these or similar events and trends were isolated or remote so we need not be disturbed by them. No doubt many are praying - regardless of whether they are willing to admit that that is what they are doing - that they will still not be affected.

Today, we <u>are</u> being disturbed, threatened, and harmed by these happenings. Who can understand why they are happening? Can you hide from the conclusion that something has gone wrong? Have we lost control of our society? More questions are being posed about religious beliefs. Some have abandoned the beliefs they were taught and adopted other beliefs. Some ask, "Is religion part of the cause?" Is there a "right" path?

Choice of Worlds

Why are these pictures important? Because they become a permanent part of us. They are the raw data from which we form the bases for our rules about we conduct ourselves, plan our lives, and interact with others.

What picture do <u>you</u> have of the world you live in? What kind of a world would you prefer to live in? Would it offer:

Comfort? No hunger, no pain, adequate shelter
Security? No wild animals, no enemies, no natural catastrophes
Freedom? To act on your impulses to do what you want to do when
  you want to do it

Pleasure?   Satisfaction of all your desires.

How close are you to realizing this kind of perfect world? Suppose you could have your wish. What would life be like in such a world? Pleasant? Routine? Would you prefer the excitement of some "hardships" rather than the monotony of the perfect world?

But let's say you do not want to experience the grief of having your child eaten by some animal, or have some tribe from a couple of blocks away do you in some night while you are sleeping, or succumb prematurely to some epidemic, etc. How much influence have you had in shaping your world? Do you prefer to believe you have some control over your life? Have you reached the point where you sense that somehow it does matter?

Then you have reason to read this book. It is not a book on gloom and doom. It is a book about answers that have made sense to me. It is about challenges and directions. It is about our survival as a life form on this planet.

Chapter 2. Plan

Motive

My motive in writing this book stems from gratitude that my own little world has been reasonably peaceful, and from a realization that that "peaceful" world is a heritage from millions of butchered bodies. Perhaps you share such a feeling. Perhaps you can be persuaded that we must join in preserving the world in which we are striving to live and prosper.

Whatever may be said for the utility of violent conflict, it is at least wasteful. Among the conflicts are those about religious beliefs. This book focusses on the one between Science and Religion. I believe it is possible and necessary to reconcile those who deny a God exists and those who hold He lives.

Purpose

The ultimate objective is the total cooperation of all in achieving a utilization of our physical and human resources that at least minimizes the pains and ills and misfortunes that afflict us.

The initial purpose is to challenge you to think about the human condition and to help define the conditions under which a) peaceful and productive coexistence, b) joyous living, and c) long term survival are possible. I suggest that realizing these conditions amounts to achieving the maximum possible degree of human freedom.

Approach

We begin by extracting from our accumulated knowledge essential elements from which to build a consistent, comprehensible, credible belief system for the world we have inherited. The bits of information herein are offered for your review. I suggest a synthesis that makes sense to me to serve as a starting point. References are given for important facts and ideas.

You will recognize many familiar ideas in this synthesis, but you may not have believed that they can be organized to point out a direction for our efforts and a set of goals. Provided we discard ineffective ideas; formulate the ideas we retain in a readily understandable and readily believable form; and fully understand how much our well-being depends on these ideas.

Scope

This review will touch on many areas of thought. We need to identify requirements for our information.

A. Facts

We must be able to distinguish between facts and assertions, information and interpretations. I shall use dictionary definitions where possible so that you can easily check them, but I shall use specialized meanings where appropriate.

- "Facts" are verifiable observations. E.g., we all accept the Law of Gravity, because we can check it out for ourselves, even though we cannot explain it.

- An "assertion" is a statement made boldly, usually without other proof than personal authority or conviction. E.g., the Earth is flat.

- "Information" is a collection of facts, theories, and hypotheses, whether or not they are related.

- "Interpretation" is the meaning given to something by someone, sometimes involving personal opinion.

B. Beliefs

There must be a mutually acceptable set of beliefs, because beliefs underlie our "explanations" for what we observe and our decisions on how we conduct our lives. Let me give some examples.

- In the physical world:

  We won't have a (flood) (tornado) (earthquake) (volcanic eruption) because we've never had one.

  It is impossible to construct a ship out of iron, travel faster than (8 miles per hour) (Mach 1), or fly through the air.

- In the social world:

  Man (i.e, men and women) is (good)(evil).

  Women should be (submissive to)(equal to) (dominant over) men.

- In the scientific world:

  There is no limit to discovery by Science.

  Science must be free to conduct <u>any</u> experiment.

- In the religious world:

  We all must ultimately answer to our Creator.

  There is no God; it is all a fairy tale.

Question: How did we come to hold the diverse perceptions, opinions, and convictions reflected in our multiplicity of religions, sects, cults and other organizations? How did our values change?

We need to look critically at our information and our beliefs. A serious problem is that increasing numbers of "educated" people cannot understand the meaning and implications of written material, or think logically.

Question: How do we choose our beliefs? Can a coherent society accommodate more than one set of beliefs about a given subject at a time?

But whatever beliefs we agree on, we must be willing to believe them with all our hearts and souls, to practice them, and pass them on to our children.

C. Values

Ultimately, there must be a practical system of values. What are values? "Moral values" refer to the accepted customs of conduct in a society. In ethics, a value is a quality desirable as a means or an end in itself.

The distinction between ethics and morals is not clear. They are even listed as synonyms. Morals seems to be the rules of right and wrong conduct between individuals, while ethics deals with the rules in business and professional matters. Compliance means following both the letter and the spirit of the rules. Both sets of values may be different in different societies and can change with time. An example in ethics is that if you make a verbal contract you will honor your word even if it turns out unprofitable for you to do so.

A value is thus some standard of behavior by which we can judge how well the available options for action meet that standard. Many values are polarized with gradations: trustworthy/treacherous; honest/deceitful; brave/cowardly.

Neither ethical nor moral values seem to have had a universal and immediate <u>compelling</u> force. So we have devised ingeniously written contracts and an ever more complex and costly system of laws, but find that we are equally ingenious at getting around them.

D. Goals

There must a consensus on goals: what is the ultimate end of human actions? Some examples are:

- Survival of the human species
- Improvement in the quality of life

    o Minimizing toil and hardship
    o Experiencing pleasure
    o Living without violent conflict with others

- Defining a meaning for life

Since we are living in a time when many of the old "values" are being abandoned, logically we must not prematurely conclude that the following are not valid:

- Living in violent conflict with others
- Emphasizing the self over the interests of others
- Inflicting pain on others

But if we should conclude these are valid, we must be prepared to be the receivers of the pain they cause. Perhaps we should reflect on:

- **Hamlet, Act II, Scene 2 (FQ, p 261a):** "There is nothing either good or bad, but thinking makes it so. . .": and,

- **Isaiah 6:21:** "Woe unto them that call evil good, and good evil."

Comment: Many of us have heard the Shakespeare quotation. So it is buried in our subconscious minds. But how many have seen it as insidiously introducing what became the nihilist philosophy of three centuries later?

E. Methods

Finally, in resolving our conflicts, we must agree on methods for changing our condition for the better. There is a gamut of conflict resolution techniques:

- I   am bigger/stronger   than you.
- My father   is bigger/stronger   than yours.
- My tribe   is bigger/stronger   than yours.
- My country   is bigger/stronger   than yours.
- My God(s)   is/are bigger/stronger   than yours.

With the advent of nuclear power the system called MAD (Mutually Assured Destruction) emerged. It said you may vaporize us, but before you get us all we will vaporize you. There is hope, however, because the major powers accepted this doctrine and backed off from a nuclear showdown. We must not forget that MAD remains a perpetual fearful reality to us all and especially to those who would let down their guard.

As to peaceful methods of conflict resolution, we have two approaches: faith and reason. "Faith" is based on believing what our ancestors taught us plus what we are willing to accept without "scientific" proof. "Reason" says yes, but we must also include what we can find out about our world. Each has a fundamental assumption.

- Reason: It is possible to observe phenomena accurately and to construct an internally consistent explanation for them.

- Faith: The internally consistent explanation for our existence is in the (Bible)(Koran)(Tanach) (etc.). Tanach is the Hebrew Old Testament.

The faithful say these are revelations from God, not assumptions. Others may examine what our ancestors passed on to us in these books and say that this legacy rests on their "reasonable" interpretations.

When we dig into the arguments, we find that neither is pure. Both approaches have "facts", assumptions, interpretations, and unanswered questions. Some insist the Bible is infallible. All right. Which Bible?

- The Tanach.
- The Septuagint, the oldest Greek version of the Old Testament, about 300 B.C.
- The Tyndale Bible of 1525, the first translation into English.

- The Douay Bible, an English translation by Roman Catholic scholars, 1610.
- Seven Protestant versions later: the King James Version (**KJV**) of 1611.
- We now have the American Standard Version (**ASV**), the Revised Standard Version (**RSV**), and the New International Version (**NIV**).

Comment: This despite the command of God:

> **Deuteronomy 12:32:** "What thing soever I command you, observe to do it: thou shalt not add thereto, nor diminish from it."

And the threats in **Revelation 22:18,19**:

> ". . .If any man shall add unto these things, God shall add unto him the plagues that are written in this book: and if any man shall take away from the words of the book of this prophecy, God shall take away his part out of the book of life. . . ."

It is claimed that (1) the earlier versions contained errors as judged by the text of newly discovered but more ancient texts and (2) the changes did not change any of the major teachings of the Bible. Perhaps. But at least subtle changes are creeping in. Are the following from **Isaiah 45: 7** equivalent?:

KJV "I make peace, and create evil: I the LORD do all these things."

NIV "I bring prosperity and create disaster; I, the LORD, do all these things."

Comment: "Disaster" brings to mind earthquakes, floods, tornados, etc. "Evil" evokes images of lying, theft, adultery and murder but also disease, accidents, war, etc. "Prosperity" deals with material wealth whereas "peace" deals with relationships.

It is also strange that believers in an omnipotent Creator God allow themselves to be drawn into debates that can only proceed on principles of reason rather than faith. After all, such a God can change the Laws of Nature at will, as He may have done as recorded in **Joshua 10:12-14** where He granted Joshua's petition that the Sun and Moon stand still for a day in their orbits. He can also cause miracles, that is, events that cannot be explained in human terms.

17

Charles H. Peterson

For this review, we can proceed only if we set aside such considerations temporarily. What methods do we have for our search? A specific rational method is the principle of cause and effect. This means:

- An effect E is caused by cause C alone; and
- Cause C can result in only effect E.

Unfortunately, our Universe is not that cooperative, and our knowledge of it is far from complete. For:

- Effect E may be caused by:

  - One or more other causes operating alone; or
  - Some combination of two or more causes; or
  - A so far undiscovered cause.

- Cause C may result in:

  - One or more other effects individually; or
  - Some combination of two or more such effects; or
  - A so far undiscovered effect.

However, we cannot think of <u>all</u> the possible questions. It is instructive to recall a fantastic proof by Kurt Goedel, a mathematician, in 1931 (**TWM.** Goedel's Proof). His conclusions are in mathematical language, which might as well be Sanskrit to most of us:

1) No meta-mathematical proof is possible for the formal consistency of a system comprehensive enough to contain the whole of arithmetic; and

2) Any system within which arithmetic can be developed is essentially incomplete.

I believe these can be translated as follows:

1) It is impossible to prove logically that our system of mathematics is internally consistent; and

2) Even if our system of mathematics is internally consistent, there are true arithmetic statements which cannot be derived from that system.

Science can only show an external plausibility: a new finding must not contradict or violate anything that has been "proven" (read: accepted) thus far. Religion has a difficult task since it insists the words in its Holy Books are infallible. Any change in <u>any</u> of the words tends to undermine the authority of the Books.

Science has a great advantage over Religion: it will revise or abandon <u>all</u> its previous conceptions if they are shown to be incomplete or incorrect "explanations" (read: descriptions) of our Universe. E.g., Newtonian mechanics is now seen as a special case of Relativistic mechanics. Yet Science is limited in that new propositions may have to come from a mysterious "insight" or "intuition".

An important method of checking our findings is to repeat the observation. However, this takes care of only procedural and arithmetic errors. More convincing is an independent method of checking them.

Another method is common sense, which is the use of normal native intelligence and sound practical judgment based more on experience than on specialized training. Unfortunately, common sense has gotten muddied up with perceptions and ideas that are inaccurate and even incorrect. We ask if a given result is "reasonable", does it conform to our expectations? The problem is when and how do we know it is giving us correct answers or guidance?

Through it all, let us retain a degree of skepticism of all claims, be they of Religion or of Science.

F.   Basic Questions

What you have read so far was intended to encourage you to let your mind sweep over images of mundane objectives, nonpersonal goals, hardships, conflict, and survival. Day after day these images are with us. We need to sort out the images, understand them, discern where they lead us. And so eventually we are led at least occasionally to ponder ultimate questions. There can be no more ultimate question than:

Is there a God, or did our Universe exist forever?

Chapter 3. Perspectives

Short Range View

We can proceed a day at a time and hope that somehow we will survive. Many do just this. Humans have been around for a long time. Some think it is a matter of some 6000 years; others say one or two million years. Still others say it is nothing compared to dinosaurs. Or cockroaches. Or ants. Or bacteria.

Even in our shorter history, there have been many changes. Empires have come and gone, some without a trace. The ordinary human has always been an expendable module. But even rulers are expendable. If we shrink from this image of ourselves, we need to take some time to understand where we are, what direction we are going in, and what can we do about it.

A. Physical world

Let me offer a practical example in the physical world. Our oceans are vast compared to us tiny humans: they are measured in cubic miles. Today large populations of the world are dependent on millions of tons of food taken from the oceans. But we are dumping millions of tons of wastes from our "civilization" every year into them. Some of our "civilized" nations have been dumping highly radioactive wastes. Some of these are being biodegraded, or attenuated through natural decay processes. Where the dumping rate exceeds the decay rate, the concentrations are building up.

But we are getting a warning. Many species of frogs all over the world are disappearing. We see disoriented whales. Why? Is there some effect from some chemical we have put into the environment? Many substances in minute amounts, like parts per billion, have profound, even lethal, effects on us humans.

> **Illustration:** An everyday example is prescription drugs, doses for which are often in milligram quantities per pill. One mg in a 70 kilogram (154 lb.) human is one part in 70 million. Another example **(The Wall Street Journal**, 14 August 1995, p B1) is that many are allergic to peanuts. A 1 mg amount is enough to cause illness. The allergic reaction can include throat constriction, stomach cramps, wheezing, shock, cardiac arrest, and death.

Illustration: Consider vitamins. The Recommended Daily Allowance of Vitamin B-12 is 6 micrograms. This is about 1 part per 12 billion of body weight. It is needed in all cells and for synthesis of DNA, RNA, proteins, nerve cells, etc. Deficiency shows up as mood swings, memory loss, dementia, degeneration of nerves, and anemia.

One part per 12 billion is 1 microgram in 12 liters (about 3 gallons) of water. How much is a microgram? Well, the period at the end of this sentence is a cylindrical dot about ½ of a millimeter in diameter. Suppose it is a tenth of a millimeter high. Its volume is then 0.0065 cubic millimeters. Its density is probably like that of water: 1 mg/cubic millimeter. So the dot weighs about 6.5 micrograms. Dispersing it in 21 gallons of water gives us 1 part per 12 billion.

How can such tiny amounts have any effect? Because a gram has a great many molecules. Suppose the contaminant has a molecular weight of 200.

Explanation: A molecule is the smallest quantity of a substance that has a definite formula that still has the properties of that substance. Carbon-12 is the basis for a relative weight system for atoms. It is set at 12. The units can be grams, pounds, or tons as long as one is consistent in the calculations. The sum of the weights of the atoms in a molecule is its molecular, or formula, weight.

For every such substance the number of molecules in one gram formula weight is the same: 6.02E23, which rounded means 6 followed by 23 zeros. One microgram contains (1E−06 g/200)(6E23), or 3 quadrillion molecules. At 203 teaspoons/liter, every teaspoon of water would contain about 1.2 trillion molecules.

These may not have a gross effect, merely disturbing some complex equilibrium among the life forms in the oceans. At some level of pollutants life will not be possible in the oceans because of direct toxicity, or elimination of some necessary life form in the food chain, or interference with reproduction. Some lakes in America already have enough mercury in them that although we can drink the water we can no longer safely eat the fish from them due to bioaccumulation. How many tons of pollutants can we put in the oceans?

Illustration: To make a rough estimate, figure the Earth as a sphere with a mean radius of 6.38E08 cm (3963 miles). The oceans cover 3/4 of the Earth's surface and have an average depth of 12,000 feet, so the volume of the sea water is 1.40E24 cm^3. Its average density is 1.028

g/cm^3 and it holds 3.5% by weight dissolved salts. The mass of the water is thus is 1.39E18 metric tons.

One part per 12 billion is 116 million metric tons of pollutants that could be added if we wish to limit the final concentration to that of the example above of Vitamin B-12. This is TOTAL. If the pollutant does not degrade, no more should be added. Annually, we are adding amounts estimated in metric tons as follows (**TGR**, pp 310-313):

Ocean litter	6 million
Petroleum (all sources)	6 million

This is unfair to petroleum because there are also megatons of fertilizers, animal wastes, pesticides, industrial chemicals, household chemicals and soaps, etc. that are not mentioned in these figures that get washed down to the oceans via the rivers.

The conclusion is not that our oceans can accommodate only nine years worth of such pollution but that we are uncomfortably close to converting them into a vast lifeless garbage dump.

> **Comment:** This is, of course, one answer to the exploding population of the Earth: cut off the food supply and the excess will simply starve to death.

A serious problem is that what we dump is not immediately dispersed over the entire ocean. Thus, some sectors of the ocean are immediately rendered uninhabitable because of oxygen depletion. So fish and other aquatic life suffocate.

B. Social world

Examples in the social world are more complex because many factors interact. And they operate over longer time spans. It should be sufficient to point to the proliferation of the will to use violence to solve human conflicts as an indication of the inadequacy of our institutions, presaging their eventual collapse.

When does it become an emergency? When must we change the pattern of our familiar, comfortable ways? When do we consider what are the possible solutions? Or do we prefer to behave as if we are on a conveyor belt through life and dare not ask where it is taking us. . .?

We have gained many things by crowding into cities but we have also lost much. E.g., those who still live on farms learn early what a dark cloud on the horizon may mean. Experience shows that sometimes it never gets to where they are, but when it does it often means a storm, perhaps a tornado. Those who stayed within reach of the German authorities in the 1930s because they chose to believe the dark cloud would never reach them paid the ultimate price for not considering that things might be different: they died after much pain and agony. The unthinkable, the impossible, the improbable happened.

And there are many other examples of the consequences of wishful thinking: Russia, Poland, Manchuria, China, Korea, Cambodia, Tibet, Cuba, Pakistan, Afghanistan, Yugoslavia and many countries of Africa.

Current Situation

All of the negative trends, and even some of the positive trends, noted in Chapter 1 are challenges to our survival. All have been developing for centuries, manifested in different forms depending on the technology, the understanding, and the power structure of the times. Somehow we have muddled through. For our present challenges, there is no consensus on a comprehensive plan of attack on them. I am not aware of any group or power effectively addressing these challenges and leading us toward a saner world. Thousands are devoting their energies, resources, and lives to spreading their various forms of religious and other beliefs. But I have the uneasy feeling that we are losing the struggle for the good life.

I see individuals and groups striving for their own gratification, enrichment and survival, regardless of whether any significant number of others survives and prospers. Some argue for preservation of old institutions; others say we must do away with them. We have confusion, contradiction, and conflict instead of clarity, consent and cooperation.

Challenges

We need to get more people thinking about our situation. . .How we have become what we have become? Why do we think what we think? Why we do what we do?

We have heard these questions before.

Romans 7:15 ". . .for what I (Paul) would, that do I not; but what I hate, that do I."

23

We need to identify what our civilization rests upon. Picture the bridge arching across a river. But think also of the unseen foundations buried in the earth without which the bridge could not stand. Think of the contributions of myriads of people doing their daily tasks and of the many agreements on how to conduct ourselves that we have developed over the ages.

No one can read everything, understand everything, think of every situation and contingency. We need to have better interchange of ideas. We need to identify what is possible. We need to find out what it is that we can agree on, not just what we disagree about. We can accomplish nothing but our own destruction if we only disagree. We already agree on many things.

- We agree to use a common language within a given society. Why? Because it facilitates innumerable interactions within the society. How can we develop relationships if I jabber at you in English, and you jabber at me in Mandarin, Hindi, or Arabic?

- We agree to observe particular restraints on our behavior, like customs, rituals, and laws. Why? Because they minimize conflict, help make social interactions more enjoyable, and make the society more coherent and productive. And because these make freedom possible. . . .

We must strive for understanding that unites us in peaceful productive cooperation rather than in destructive conflict. This is the path to experiencing the thrill of life, rather than spending our energies on attack and defense, fear and anxiety, destruction and death.

Chapter 4. On Information

Questions

As the innocence of childhood passes, questions emerge to which we can get no answers. There are good times, but also periodic upsets. As fast as we climb the ladder of success, for some the rungs keep breaking. For others, the ladder even seems to be hidden. Then there is the disillusionment and pain of broken marriages, serious illnesses, wayward children, etc. We cry out for answers, explanations, reasons to continue.

We learn that even our mentors do not have answers, and often make decisions based on some concept that they have come to believe to be true. They also tell us we can cope with and even solve our problems if we believe ideas that we find are difficult to reconcile with other information we have. So we have developed a variety of responses to questions.

- Eliminate the question.

  Ignore it: Don't even acknowledge the question.

  Suppress it: Shout it down; ridicule it.

  Deny its importance: Their purchase of materials for a nuclear bomb is unimportant since they can't deliver it.

  Deny its urgency: It will take years for them to build a bomb.

  Deny the possibility of answers: There is no way we can prove they can build a nuclear bomb.

  Deny responsibility for answering it: Our department collects information, but does not analyze its implications.

  Deny ability to answer it: We just do not have the facilities or know-how to deal with it.

- Eliminate the questioner.

  Send him away: This is a short term expedient.

Put the questioner in jail or in an asylum: This is a longer term expedient.

Kill the questioner: This was useful when a subject dared to approach his ruler saying, "O mighty ruler, how is it that thou art ruler and we are subjects?"

- Eliminate ourselves.

Escape to another location: Get transferred, or change jobs.

Escape into another world (alcohol, drugs, fantasy, insanity).

Die: End life by natural death; acquire an illness that shortens our lives; commit suicide.

- Answer the question.

All except the last response amount to avoidance. It is true, though, that in a busy world the first seven are useful screens to protect ourselves from involvement in too many problems. But avoidance here means finding excuses for refusing to give some of our time, energy and money to help deal with life's persistent serious questions. We fill our lives with work, play, eat, study, rest, and sleep. We go on many days without seemingly being affected by these questions. We get to hope we never have to answer them. But one day they can no longer be ignored.

- Like a leaky faucet that has now corroded and cannot be repaired by simply replacing a gasket.

- A field that after many years of irrigation is too salty to bear crops.

- A flirtation that destroys first trust and then a family.

There is another reason why we generally do not think about our condition. Our minds and souls are overloaded with stimuli: there is no time for reflection. Newspapers, magazines, the movies, radio and television beckon, seduce, and saturate our minds with messages:

- You are nothing unless you buy our product X;

- Everybody is enjoying the good life except you; that is because you haven't bought our product X;

- You have a right to lust after any passing body that catches your eye.

A current advertising technique is to flash a sequence of images so rapidly past our eyes that before we have had time to think of the implications of one image we have been confronted with a dozen others. To further catch our attention, the background sound track is filled with a loud cacophony of "music" that often involves the pounding distraction of a drumbeat. Its effect is to interfere with our intellectual processing of the information offered. We seem to have an inherent tendency to respond to rhythm, so that helps insinuate the message into our consciousness.

The last option is a struggle to figure out what is going on. We do sense that the questions are important. We cannot wait until the problem represented by the question presents an unavoidable serious threat to our survival, not just to our peace of mind.

"Solutions"

Many "advisers" have proposed "solutions". Why don't we use them? Initially we do. For awhile. But if a "solution" doesn't improve our condition soon, we lose faith in it and turn to another. Most of us have no time to study them in depth and evaluate them.

We notice, however, a few things about "solutions" that make us suspicious of them. Some of our "advisers" give fancy names to events and situations, and talk a great deal describing the problem but do not actually come up with solutions. Some require actions directly opposite to those of others. How can they all be "correct"? Many "solutions" fail because they treat only part of the problem, as well as ignore information which does not fit the assumptions of the solution proposed. We find much of what we are told is irrelevant, insincere, or ineffectual.

We have also learned that we must be very careful as to whom we give power over our minds and bodies. So it seems inescapable that we have to dig in and find out how to choose. It may be that all "solutions" are temporary. They are devised to fit the then current set of conditions.

- People used to seek safety in fortified towns. Even these could be captured by ordinary weapons, or by starving the people into submission. With gunpowder, they could be destroyed from afar.

- We strive to stay one step ahead of microorganisms with our cough syrups, aspirin, penicillins, cyclosporins, and now gene therapies.

27

- Is the problem of feeding ten billion people generically different from that of feeding a family?

Sources of Information and Ideas

We can reflect on our own observations, or talk to other humans. With libraries, our information horizon mushrooms. An astronomical number of hours of study over the centuries has already been invested. No one has the time just to turn the pages of but a tiny fraction of the written records of that toil, even if we could understand the languages in which they have been written. Words have changing meanings. Some are not even known to later generations. And we do have full-time jobs and families and other responsibilities.

We must rely heavily on what scholars have found in their studies. We have conducted too many thousands of years of experiments in human relations to say the information from them is useless and should be discarded. We have too much information about our Universe and ourselves to ignore the inconsistencies, errors, and obscurities in what Science asks us to believe.

A serious obstacle is the fact that much of modern Science is incomprehensible to those of us who are not trained in the various specialties. Some writers have offered translations from the esoteric. Desperately needed is a further level of translation so that more of us can check out for ourselves what is claimed and asserted as truth.

We must also add our own perceptions, ideas, and interpretations because we have learned that no one human has all the truth. We learned a long time ago that we cannot put our full trust in any human and all of us have our own interests at heart.

> **Psalms 146:3** "Put not your trust in princes, nor in the son of man, in whom there is no help."

There is also another strange source. I can't explain it, but many times thoughts have surfaced in my mind on which I was unaware of any prior thinking or reading. E.g., in my early years, I was aware of thinking that we humans did not have any purpose in life. We were not born with a label stating any purpose. But immediately the thought came that those who assumed a purpose for themselves seemed happier. I had no reason to believe it was true, but it made sense.

When I read Omar Khayyám's advice (**TGP**):

Yesterday <u>This</u> Day's Madness did prepare;
   Tomorrow's Silence, Triumph, or Despair:
Drink for you know not whence you came, nor why;
   Drink for you know not why you go, or where. . .,

I rejected it, though not through any long process of analysis. Somehow I wanted to believe that life meant more than drinking wine.

In my search for answers, I found many friends. Most of them are dead. I mention only a few of the more familiar ones, though not in any order of preference, and certainly not implying that I agree with all that these have written. I found inspiration, intellectual stimulation, and comfort in these friends, as well as several false trails.

Philosophy	The Greeks, Schopenhauer, Russell
Science	Galileo, Newton, Mendel, Gamow
History	Durant, Toynbee
Religion	The Bible, Campbell, Frazer
Literature	Shakespeare, Bullfinch, Fitzgerald, Carroll, Kipling

These and many others have given us ideas to consider in our efforts to understand our condition. Not all their ideas were valid or are applicable today. You can decide for yourself as to their validity.

Much of what I say has probably been said or thought of before. Perhaps a thousand times. By people I have never met and by many that I can never know because they left no record. My contribution is to put these ideas together to help answer our basic question.

Eclectic Pragmatism

In the expectation that we can extract some ideas and principles from all that has been said and written and tried in the past to guide our lives, I propose Eclectic Pragmatism (EP). Its meaning is as follows:

- **Eclectic:** Drawing upon knowledge, wisdom and truth from whatever source.

  1 Thessalonians 5:21 "Prove all things: hold fast to that which is good."

- **Pragmatism:** Stressing practical consequences as constituting the essential criterion in determining meaning, truth, or value.

Translation: The meaning, truth, or value of an idea or act cannot be determined without considering what follows from the idea or act.

Comment: A serious obstacle is that consequences may take decades to develop, be recognized, and understood. Note especially that the definition is not restricted to <u>material</u> consequences.

Why be eclectic?

- We have already been eclectic in our beliefs.

    o   Our religions include elements from a wide variety of beliefs and practices.

    o   The Bible itself is the result of choices by unknown humans as to which writings to include.

    o   Preachers choose which parts to emphasize and which to ignore. How many sermons have we heard on Leviticus? Ezekiel? The Song of Solomon?

- We have to be eclectic.

    o   In school, there isn't time to teach all that went before. The facts and ideas which we present to our children must be selected and edited.

    o   Changes in our total environment are changing the relevance of our information.

Some of these facts and ideas are critically important. Who is selecting what our children are exposed to and are pressured to incorporate into their thinking?

Why be pragmatic?

- We need methods and ideas that meet our needs.

    The principle of pragmatism often does not require much argumentation in the physical world. You don't walk around in the Arctic in only a loin cloth. It is in the world of beliefs that thorough examination and discussion are needed.

- There is a time element in our problems.

  We are concerned with our immediate survival, so we often are forced to adopt a course of action before we have had time to investigate it theoretically or philosophically.

EP involves a statistical viewpoint. In the material world, some hypotheses have the status of laws (being near the 100% true end of the scale) while others are rejected (being at the 0% true end). Intermediate in status are theories, surmises, and speculations. In human relations, the easy approach is to view everyone as alike, even though we are urged to see ourselves as unique individuals. Maybe this means each of us is to think of himself/herself as unique, but everyone else is a clone, a blob. EP recognizes us as different - physically, mentally, and spiritually.

By a survey, we can determine the distribution of each characteristic. E.g., perhaps less than 0.00001% can run faster than 15 miles per hour. Maybe 0.01% cannot move at all. Perhaps 10% can make their bodies move faster than 3 mph. The rest are in between 0 and 3 mph. The distribution here is continuous: mobility changes gradually over the whole range. The exact distribution is not as important as recognizing that these differences are real and the solution for any problem involving that characteristic must consider the range of that distribution.

The ideas in this book are intended as an initial basis because they make the most sense to me now. Readers should examine them to see what they can agree with in the effort to enlarge the area of consensus.

Facts

We have accumulated much information from experiences, observations, and contemplation. Humans once feared thunder and lightning, believing they either were gods or were the evidences of angry gods. We now have material explanations for these phenomena. We have quantified many of them, meaning we have developed laws that let us predict the behavior of materials and energy.

Over the centuries we have slowly reduced the sphere of the unknown. The questions remaining are the difficult ones. When we die, does some part of us survive? If so, where does it go? What is that place like?

It appears helpful to start with "facts". We need to agree on what we can observe about our Universe by whatever sensors are available to us. We can look for patterns and trends in these "facts" that lead to "Natural Laws". These are the What, the Where, the How, and the When of our Universe. Then we can move on to less observable and demonstrable considerations. One is the Who. Another is the Why.

We need a common data base because in the past, and even today, people have been taught and coerced to believe unreasonable and incorrect things about our Earth, our Universe, ourselves, and our society.

Authorities

Other questions are less readily settled:

- Why should I care about others?;
- Why do we put so many restraints on human sex?;
- Where do we come from?

For answers, we turn to "authorities": sources of information whom we believe we can trust to be knowledgeable and truthful. Initially, these are parents or other older family members. Then it is teachers. Beyond them is the police force and ultimately the nation's rulers. Their common characteristic is that they can use threats of punishment and actual physical force to make us comply.

But there are also intangible authorities like tradition and religion. When we can't answer a question it is convenient to say, "Because my God made it that way". It is more convincing than "That's just the way it is." But why should I accept the teachings of your God rather than the one I was taught to believe in?

Truth

Answers to the questions raised in this book are difficult to formulate because of:

- Conflicts of individual desires
- Many aspects - physical, economic, political, historical, religious
- Differences in information, beliefs, and desires.
- Unwillingness to accept the need for change.

However, we must make the effort to find answers to how we are going to conduct our lives every day because others are already making decisions affecting the world we are trying to survive in. We should start by agreeing on what information is acceptable for our inquiry. We quickly run into the problem of what is "truth".

A "truth" is a verified or indisputable fact, e.g., the Law of Gravity. If I release some object like a book from my grasp, such objects <u>always</u> fall towards the ground. No exceptions. Thus, facts are a class of truths known and verifiable from experience or observation. All who repeat the event generating the fact can see it for themselves. For a one-time event, the truth may never be known.

But we are not finished with "truth". Dictionaries generally agree on definitions, so a source is given here only when a different nuance of meaning is used.

"Truth": conformity with reality.

"Reality": a thing exists and is not just an idea.

"Exist": to have actual being.

"Being": in philosophy, that which has actuality either materially or in an idea.

"Actuality": reality.

This is circular and uninformative. For material things, we use common sense. Thus, rocks exist (have reality). They are hard, heavy and impenetrable.

But ideas can also be thought of as real. We can think of a machine that can fly through the air. The machine may not yet have been invented, but the idea is a specific thought. Initially it is in our heads, but we can describe at least its characteristics on paper. So we can say the idea is real but the machine is not since it has not been actualized, made material.

What are the criteria for reality? We might say that an idea has reality if it has a structure, or can be used to predict or explain something. Examples are beauty, justice, evil, etc. The things the ideas deal with may not exist themselves but the ideas may be useful in human relationships.

What about the idea of God? We may use many of the characteristics ascribed to God in human relations. So the idea has reality. But is God Himself real?

33

Nature of Proof

A. Repeatability

"Proof" in Science involves being able to repeat an experiment. Without repetition, such a "proof" is merely a plausible "explanation".

If I had an accident, someone may say that an evil spirit afflicted me. Unless such a spirit appears to other observers, either in some form or by some other acts, many today would reject this explanation. If I had a series of mishaps, people are still more apt to say I am unlucky than I am cursed by an evil spirit. Yet the subject matter of this inquiry is such that it would be premature to dismiss this possibility.

"Faith" is a belief not based on proof. This implies that some beliefs are based on proof. A "belief" is defined as confidence in the truth or existence of something not immediately susceptible to "rigorous" proof. It is thus something a bit closer to fact than faith. Usually there are at least some reasons for holding a belief. I would rather define faith as an acceptance of something as true independently of proof. For "proof", we find we are on shaky ground.

"Proof" is evidence sufficient to establish a thing as true or believable! This sounds circular. When is the evidence sufficient? "Rigorous?", meaning logically valid. "Logical" means by the principles of logic.

We have two approaches. We can merely think about a question, but make no observations. Or we can make observations and measurements, and think about them. Over the centuries we have found that we can make a greater amount of sense and create a more permanent amount of order, at least in our material lives, by including observations. Even religions have done this, e.g., by devising rituals related to the fact that seasons do recur on an annual basis and grass and green leaves and flowers do come back in the spring though apparently dead in the winter. Science still doesn't know how this came about.

We really cannot conclusively prove many things. But a principle based on repeated trials is that we accept something as true if we have found it is true always and never found it is false. A single exception is enough to cast doubt on a hypothesis (a proposed explanation) or cause its rejection. E.g., generation after generation, we observe the cycle of seasons, with never an exception. So we live as if it will always be so.

Comment: The corollary is we reject something as false if we have always found it to be false and never found it to be true. E.g., Evolution says life forms change through transition forms. None have ever been found and every fossil found fails to show evidence of transitions. Therefore, Evolution is false. Yet many cling to it fanatically.

B. Consistency

There is another aspect to proof. Through analysis of our observations, we build a body of facts or truths that is systematically arranged to show the operation of Natural Laws in specific areas of investigation. These facts or truths must be internally consistent.

There are other explanations that are sometimes true and sometimes don't seem to be true or to apply. Here it may be that there are other factors operating that we are not aware of and thus are unable to allow for them in our hypotheses.

C. Usefulness

Another principle is usefulness: if the first conclusion is true, then we can usually think of others that must also be true, that "logically" follow from the first. This is illustrated by theorems of geometry.

Illustration: We agree on what we mean by straight lines and circles. Then we agree on some "obvious", "common sense" rules, like two straight lines can intersect in one and only one point. From such humble beginnings, observations and reasoning about straight lines, parallel lines, and circles lead to formulas for calculating areas and volumes of geometric figures, distances to inaccessible objects, and much more.

D. Logic

We have found two methods to investigate our Universe.

- Observation. If we had no sensory capacity at all, we could not tell if there was a Universe. However, we have also found that we cannot be sure we are interpreting our sensory data correctly.

- Inference. We can extend the range of the knowledge gotten by observation by reasoning. E.g., to explain day and night, since the only light source early humans knew was the Sun, it must be travelling around the Earth. Later, a second possibility was conceived: the Earth could be rotating.

In addition we need rules for logic. Our inferences must lead to consistent conclusions. If we are counting coins, candles, or cabbages, one and one always must be two. However, one and one could be six billion, since the Creationists tell us our present world population came from an original pair of humans. So we must define our terms carefully, and somehow derive and agree upon principles of how to reason.

Thought vs Observation

We need to consider another question: why do we think we are capable of explaining our Universe? One reason is experience shows we <u>are</u> capable of explaining many things. Thus, we can anticipate problems, predict events, and devise timely courses of action. Why? To survive. To satisfy an inner drive.

> **Question**: Where does the drive come from? It is no more sufficient to say it just is than to say God just is.

The drive to explain goes further. We have curiosity: we seem to reach out for knowledge and understanding for its own sake. It may be a part of the drive to survive by investigating the different, the new. It also has a component of ego satisfaction in demonstrating one's powers, if only to oneself.

However, it is an assumption for Science to say the principles by which it measures things on Earth can be applied anywhere in the Universe.

> **Comment:** In Religion, this approach is called faith.

Some call it a hypothesis, and then forget that it is just an assumption. They justify it on the grounds of simplicity and consistency, but they mean without it Science could never know anything about our Universe.

If we never find exceptions to the proposed Natural Laws, our confidence increases that we have found at least effective, if not correct, expressions for these rules. If we can predict events successfully by them, we become even more sure of their validity. If we find some observation that doesn't fit the rules we may have to start over. "Scientific explanation" is thus practiced by a trial and error process in which new facts must be accepted even if they disprove the old. And many observations remain unexplained. . . .

Free Will vs the Will of God

Many things in Religion also cannot be explained. Advocates say we must believe that these things can only be attributed to the power of God. Our approach to life then is a choice from four attitudes.

    a. We cannot influence the will of God.
    b. We can try to influence the will of God.
    c. We are subject to inscrutable Fate.
    d. We can ignore God.

In a), everything is foreordained, but we will be rewarded in Heaven for our obedience to our perception of His will. But then the concept of obedience is meaningless: we are only puppets. We cannot be held responsible for what we do because it is the will of God. If we kill someone in exercising our will, and someone else kills us for our act, that too is the will of God. What kind of a God would be pleased to play such a game for thousands of human years with millions of humans? Another pair of quatrains from Omar Khayyám" (**TGP**) is:

> "We are no other than a moving row
>   Of Magic Shadow-shapes that come and go
> Round with the Sun-illumined Lantern held
>   In Midnight by the Master of the Show;
>
> But helpless Pieces of the Game He plays
>   Upon this Chequer-board of Nights and Days;
> Hither and thither moves, and checks, and slays,
>   And one by one back in the Closet lays."

Like a little boy playing with his toy soldiers. Or a soap opera full of violence, suffering and death, even for the believers.

The argument that we cannot influence the will of God is devastating to initiative and to personal responsibility. If everything is foreordained from before the beginning of our Universe, should I not ignore the tiger, the truck or the terrorist in my path because if it is the plan of God that I shall be harmed or killed, there is nothing I can do about it? But the wisdom of the ages says:

Matthew 4: 7 ". . .Thou shalt not tempt the Lord thy God. . ."

Charles H. Peterson

God can regard our efforts to tempt Him with amusement or annoyance. Or was it His plan for us to tempt Him? As it was His plan for Adam and Eve to disobey Him?

Question: On what basis does God decide to create millions who He knows will not submit to Him and then burn them in Hell for eternity?

In b), we must still be ready to accept that will. It ultimately prevails, because the material things of this Earth are unimportant and because we will be rewarded in Heaven for our obedience.

Question: Why should I pray?

• Who am I to ask that God's plan, which has stood for untold millions of years, be changed?

• If the plan is changed for me, then it is also changed for someone else against his will. Even if he is praying God not to change the plan. Whether I live or die, whether I pray for one career or another, one mate or another, someone else is aided or hurt or not aided or not hurt by granting my prayer and that person has not exercised any of his or her free will. That person therefore cannot have complete free will.

• Partial free will doesn't help. With free will in only 1% of my decisions, some of these will conflict with those of someone else. No, we either have complete free will or not at all. How else can we manage our lives? What is there in the results of our exercise of our wills that confirms or denies our free will?

In c), our will is unimportant because everything material is an illusion. It is as if some god were dreaming about us and if he were to stop doing this we would disappear. We cannot influence our fate and since it is illusion there is no point in trying. Eventually though we may be rewarded.

Comment: Even if we are illusions, we obviously have certain properties. E.g., we feel pleasure and we feel pain. When we are hungry, we have choices.

• We can choose to ignore the sensation because it is an illusion. Eventually we die of starvation. We eliminate ourselves from the illusory world before the Dreamer awakes.

• Or we can find something to eat even if we are only continuing the illusion.

38

So we do have some control over our fate. We can respond to stimuli. We may be a dream to the Dreamer but we are real to ourselves. Therefore, within whatever rules there are that govern creatures of illusion, I chose to consider myself as real and follow Descartes:

TWGT, p 182: Discourse on Method, Part IV (1637): "I think, hence I am."

Question: Who or what is the "I"?

In d), if His will is to affect our lives, we will cope with that as best as we can. While many aspects of our lives are fixed (we don't have wings, we need food, we are vulnerable to many calamities), we can proceed as if we had free will within our limitations.

I believe, therefore, it is a logical contradiction to say we have free will in a pre-ordained Universe. Either we have no free will in a preordained Universe or we have complete free will in a Universe that is not completely preordained. A believer can reply that our puny human minds just cannot comprehend God's plan.

Question: Then why were we given even a limited capacity to reason? To be driven to the conclusion that human reasoning is futile? To give us roughly 70 years to be beaten down in submission to God? For His pleasure? Then what is the point of thinking we were created in the image of God?

The Bible says God commanded us to believe in Him. If we choose to deny God, is that also God's plan?

Comment: We come not into this world with an operating manual. We must find our own way. I am led to choose the viewpoint which says, though my influence over my fate is small, I believe I can choose one course of action or another by an act of my will. Including choosing my God. Those who have changed religions are observed to lead different lives. Since most gods are said to have unlimited power, which god is then controlling their lives? Does the god I forsook cease to exist?

Because of all the doubts and denials of our heritage, we need to go back to fundamentals. Suppose some worldwide catastrophe destroyed all of our technology, records, libraries, etc. If we were dependent on our own abilities and resources, what could we determine about the world we live in to tell our children?

Even without a catastrophe, we need to review how reliable is the evidence for what we are told about our Universe and our origins.

- One reason is that vigorous challenges are being made to conventional wisdom.

- Another is that false leaders continually arise to take advantage of our needs to urge us to follow them to our own slaughter and their benefit.

- But the best reason is to minimize the probability of repeating the brutal, bloody butchery of our past from which we have not yet won our freedom.

We are faced with a jigsaw puzzle in which the pieces are ideas. They are mostly from the past, because we find they come into wide use only decades after someone first offered them as the New Wisdom. We find they are based on various assumptions and involve many assertions, often so worded that we do not realize they are only assertions.

Chapter 5. My Path

Each of us is on a unique path in the four dimensions of space and time. Yet I believe our paths have many similarities that we can focus on to foster cooperation. Focussing on differences erects walls. Also, knowing how a result was reached helps to understand it. So I will tell you how I came to some of my beliefs. Maybe you will recognize some similarities. . . .

Early Impressions

It seemed to me that whether or not people believed in religion, they were still self-centered and violent. Many of us are therefore growing up to say to the world, "I don't believe you or trust you anymore." I am cold and hungry. What is a Christmas present? Who is God? Where is He? If He loves me, and is All Powerful, why does He let me feel pain? What do you mean, I was born in sin? What do you mean, pain is an illusion? Leave me, I will find my own way. . . .

I had many such questions as a child. No answers. But the idea of an Eye in the sky watching simultaneously over billions of individuals was just not credible. Why would an All Powerful God be a slave to His creations over the millennia? When I learned that we live on a tiny speck of dust among billions of galaxies each with billions of stars, it was even less credible. What was the purpose of all those other galaxies if we are unique? And has God created millions of Earths? With my limited understanding of such information, I rejected God. I later found this made me an atheist, one who denies or disbelieves in the existence of a God. I was more a disbeliever than a denier.

I then came upon the idea of an agnostic. Somehow it seemed more reasonable. Such a person believed one could neither prove nor disprove the existence of God. However, I was not at ease with this fence-straddling position. Something drove me to resolve it.

My mother started me on the track to God by having me confirmed in the Lutheran faith. I did not resist, because I had respect for my parents' wishes. I was also really curious about God. I became familiar with several Bible stories and the Ten Commandments. But between full-time work in the day and school at night I had no time for serious study of religion or church attendance. My search for resolution went on hold, but I held on to some of the ideas because they seemed reasonable and useful, irrespective of whether they came from God.

Charles H. Peterson

Early View of Universe

I had a few friends with whom I debated many questions. Without schooling in astrophysics, we could still ask a simple question: Is our Universe bounded or infinite? Let me use an illustration I read of many years ago.

Most of us believe we can travel around our Earth even though we may not have actually have done it. If we don't use planes, this is travelling essentially in two dimensions on the surface of a sphere. Now suppose we were truly two dimensional people, unaware of the possibility of up and down.

> **Comment:** Bear with me. I am not a true believer in the Religion of Science. Points, lines, and surfaces do not exist alone; they are merely useful mathematical concepts. So I cannot accept this notion completely. For even a proton, the nucleus of the hydrogen atom, occupies a measurable spherical volume. So a conscious being with a thickness of even only one atom should realize that something was under it as it slid around on its sphere.

But suppose we did travel in a straight line to seek a boundary. Since we had never known up and down, it is argued that we could not perceive that our path was a curve on the surface of a sphere. Eventually we would return to our starting point.

Such a universe was described as finite, but unbounded. Today I say "finite" is understandable, but "unbounded" suggests "unlimited", which is misleading. In our travel we did not find a wall, but our two-dimensional universe was indeed bounded since it was limited to the surface of a sphere. To us, our three dimensional sphere may be contained in some larger universe to which the same question can be applied.

We can conceive of travelling away from our Earth in a straight line. Some say that space has a curvature that we can't perceive, so we could not travel in a straight line. In the 2-D case, we could not conceive of travelling along a straight line tangent to the sphere. Here we are not talking about <u>how</u> to travel in a straight line, but only about where we would get to if we could. If we are contained in some kind of bubble, we find many questions:

- Would we get to a boundary? What is its nature? How thick is it? What is it made of? Is it penetrable? What is on the other side of it?

• Is there an infinite series of worlds on the other side of the boundary? Does this notion apply to the subatomic world?

The obvious answers are a) there is a boundary, and b) there is no boundary. The problem with a) is that instead of satisfying our curiosity about our Universe it has greatly expanded our awareness of our ignorance. Here I draw upon William of Ockham, an English philosopher of the 14th century, who proposed the following principle for investigations in philosophy:

Ockham's Razor: Assumptions introduced to explain something must not be multiplied beyond necessity. More simply, the best explanation requires the fewest assumptions.

If we choose to believe there is a boundary, we have merely transferred our question from a possibly knowable Universe to a hopelessly distant and vastly more complex one. Hence, we should prefer to believe there is no boundary and our Universe goes on forever. Then we only have to explain what we find in this Universe.

But here we must take note of something I did not know in my youth. If we conduct an experiment, let's say to determine whether or not some substance has a beneficial health effect, we can make two kinds of errors.

• A Type A error occurs when we conclude that it does not, although it really does.

• A Type B error occurs when we conclude that it does, and it really does not.

Errors are, of course, undesirable, but a Type B is usually not serious because on further testing we will discover that we were wrong. But a Type A is really serious because we stop further tests of a beneficial substance! If we conclude the perceptible Universe is all there is, we may spend our resources searching in the wrong direction and thus can never find the truth. So the boundary question must remain.

There is still another possibility for our Universe: we live in an unbounded 4-D Universe that is contained in a 5-D universe that we are unlikely ever to perceive. But is that universe contained in a 6-D universe, etc.? Answer: until there is some interference in our Universe attributable to another dimension we can confine our questions to our 4-D one.

This satisfied my need to have something settled in my world. But then, where is God? As I learned more about the vastness of our Universe, God became more mysterious and remote. What I heard seemed like a pagan belief that Heaven was up and Hell was down.

- But how high is up? A few miles and we are outside Earth's atmosphere. The nearest star is 4 light-years away (about 24 million million miles).

- "Down" can only be 4000 miles. After that we are talking about up. Unless there is some kind of black hole or dimensional warp within the Earth, Hell must be up also.

- How big is God? How could He be interested in life forms such as us crawling over a speck of dust revolving about an obscure star in one of billions of galaxies?

I recall debate with a friend. He argued that everything was predetermined (which was curious since he professed no religion). I tried the random argument. Coming out of a building, at the street I could turn right or left. I would make the decision on a coin toss: heads, I go left; tails, I go right.

The rebuttal: even the coin toss was predetermined. Somehow, the energy I put into tossing the coin and my choice of what to call depended on all that went on before. This meant that we were robots operating from preset instructions to make unavoidable choices.

Now he liked to play chess. When he played, it was clear that he concentrated all his mental energies on his choice of moves. So I asked, "If you are a robot, why do you <u>want</u> to win?"

Comment: I say <u>I</u> asked. I had no idea where this thought came from. It seemed to come instantly into my consciousness. I did not spend minutes reasoning how I could respond to his argument.

He had no answer. But clearly there was something in humans we could not explain on mechanical grounds. If the will to win was not an inherent property of matter arranged as a human, then predestination was inconsistent with his scheme of things. Lacking information, we talked no more of robots and predestination.

Someone may one day prove that the will to win <u>is</u> an inherent property of matter when arranged as a human. Believers in religion can argue that the Creator of the Universe either specified this at Creation or is continuing to order our lives. Perhaps we only have the illusion of free will. And so we have another mystery: what is that part of us that is disturbed to contemplate that our free will may be only an illusion?

<u>Early Information</u>

Along the way I picked up more information, such as many diseases are caused by tiny life forms invisible to the naked eye: bacteria, fungi, viruses. I do not know if they are conscious, but it would seem they are only trying to survive too, although at our expense. They grow on the commands of their genetic codes. Our bodies also grow on the same kind of commands.

In previous ages, diseases were attributed to evil spirits. People were tortured and killed because others accused them of being in league with the devil and causing these diseases to be inflicted on their neighbors. Even today, such beliefs still exist. But why, we ask, are there bacteria, fungi, and viruses that harm us if there is a good and loving God?

I also persisted in the search for sense and meaning as I built a picture of the world I was living in. It was a tedious process of checking each bit of information against others for consistency. Other quatrains from Omar that were appealing are (**TGP**):

"Myself when young did eagerly frequent
    Doctor and Saint, and heard great argument
About it and about: but evermore
    Came out by the same Door wherein I went . . . "

"And that inverted Bowl they call the Sky,
    Whereunder crawling cooped we live and die,
Lift not your hands to It for help — for It
    As impotently moves as you or I."

These supported the notion that neither Science nor Religion had answers and that God was not up there somewhere.

I also found **"Invictus"** of William E. Henley (1849-1903) who was a cripple for more than forty years, suffering from bone tuberculosis so that one leg had to be amputated at about age 12 (**TGP**). His words supported the idea that we are responsible for our own lives and by our own efforts we can handle them. We had free will.

"Out of the night that covers me,
    Black as the Pit from pole to pole,
I thank whatever gods may be
    For my unconquerable soul . . .
I am the master of my fate:
    I am the captain of my soul."

But what is soul? I see my body; I am aware of my mind. The body is physical; the mind seems to be a manifestation of a physical brain. No one has measured the soul. An electric current flowing in a wire, produces a force around the wire that is attributed to an electromagnetic field. It disappears when the current is turned off. Does the soul disappear when the body can no longer generate the energy flow needed to carry out bodily functions and mental processes?

In 1960 Maxwell Maltz, a plastic surgeon, postulated that a human was a mind/body machine that was operated by some other element or entity (**PSY**). His observations supported the reality of something we might call the soul. Whether this something survived bodily death was not within the scope of his book.

But I was troubled by history. Nowhere in the Bible have I found any record of people being burned at the stake, or tortured, or robbed and killed by Christ. Yet all these things have been done by organized Christianity in the name of Christ. How can we forget that Christians have done these things when Christ said:

Matthew 5:44 "But I (Christ) say unto you, Love your enemies, bless them that curse you, do good to them that hate you, and pray for them which despitefully use you, and persecute you. . . ."

Matthew 22:39 ". . .Thou shalt love thy neighbor as thyself."

During the 70s, I read about Transactional Analysis, an alternative to Freud originated by Eric Berne. He and his colleagues were troubled after years of talking to patients they were still not cured. It was a kind of a poor man's psychiatry, because once one learned the basic ideas, one could practice it without the intervention of high priced professionals. We do not hear much about it anymore, although some of its methods are being used. What gripped my attention was how it got started, as reported in **IOYO**.

Wilder Penfield, a neurosurgeon at McGill University in Canada, had been working with epileptics. In one approach, he would surgically expose their brains and then probe them with a weak electric voltage to see if any particular area could be linked to the epileptic behavior. During his explorations, his patients, who were fully conscious, began remembering things, things that they couldn't recall on their own. But Penfield noticed something else. The patients were responding physiologically: their heart beat and breathing rate changed. Penfield concluded that besides remembering their past experiences: they were reliving them!

In other words, somehow our brains record permanently what we experience through our senses. Under the proper stimulus, an experience will be recalled. Thus, not only are good things passed on, but also the fears, hurts, prejudices, grudges, hates, and things called negative injunctions. These are internalized messages like "You're bad", "You are unlovable", "You can't trust anyone", "You'll never succeed", "Dislike/fear/hate/fight/kill those with other beliefs, shapes, colors, dress, manners", etc.

At this point, it struck me: "Ah, ha! So that is how the diseases of civilization are passed on to the next generation." We have these recordings within us that are like videotapes, that we are not even aware of, that influence and even determine our responses to situations. We need to become conscious of these instructions from the past if we are to live happier lives. But something was still missing, because I was unable to use the ideas in any major way in my life.

> **Comment:** Coupled with other observations and impressions I made a generalization. Many may think of education as what we get in a formal setting like school. Let me suggest that every bit of information we take in becomes our "education". It daily influences our decision making and behavior.

The Penfield experiments showed that a significant, perhaps substantial, part of our information is not at the conscious level. Consider how thoroughly we are immersed in media messages. Have you ever watched old movies? Did you notice how often the characters lit a cigarette or poured themselves a drink? What was the message? Well, right now, this is the appropriate and desirable thing to do.

I believe we are genetically programmed to at least a tendency to copy what we see around us. We are not naturally forced to copy: if the results are unpleasant, we do not copy a particular behavior any more.

Through the images and the associated circumstances, we are stirred emotionally in various ways to experience a feeling of enjoyment. The smoking and the drinking get associated with this feeling. This process is extended to other activities: driving at high speeds in expensive cars or boats; using physical force as the primary means of establishing one's point of view; winning at gambling. It is extended to ideas, although the ideas have changed over the years: climb the corporate ladder but (take care of)(sacrifice) your family; be (faithful)(adulterous) to your spouse; one hero can overcome all opponents. The technique is repeated exposure to the same subliminal message. How does this differ from programming? Or, in its malevolent connotation, brainwashing?

There was also the matter of how people who professed to be Christians were behaving. Surely such adults would support their religion vigorously and joyously. I have not seen this very often. So I allowed myself to be lured by the siren song of Science. . . .

Present Situation

Meanwhile, I have watched the civilization into which I was born being debased, disassembled, and discarded step by step. Many of the things I was urged to believe are not being practiced by the society in which I live. Consider **Exodus 20: 1-17**. Are we willing to admit to ourselves how far we have come?

1. "Thou shalt have no other gods before me".

    •   However, this seems to say we can have other gods as long as God is first.

        Exodus 22:28 "(God speaking) Thou shalt not revile the gods. . ."

    •   We have been freed from slavery but voluntarily slave away 60-80 hours a week and refuse to give just one hour of one day exclusively to God. Even in that hour, rather than getting into a worshipful attitude, we continue to babble to one another about our worldly concerns.

2. "Thou shalt not make any graven image or any likeness of anything in heaven above, or that is in the earth beneath, or that is in the water under the earth: Thou shalt not bow down thyself to them . . ."

    •   The colon may be the equivalent of "and", in which case we are given two separate commandments. We are not to make images for any purpose: no pictures of angels, no paintings, no photographs. And we are not to bow down to them. Or it may mean "because", so we can make images as long as we do not bow down to them.

• Hitler, Stalin and Mao had huge pictures of themselves hung throughout their countries. Who dared pass them disrespectfully?

• We see many two and three dimensional images around. People make them usually for money. But why do others buy and collect them? Why do they wear "good luck charms"? Do they provide comfort, security, a sense of power? "Pagans" believe that if they make an image of a living thing they can control that living thing. Some have suggested that this is the real purpose of drawings by prehistoric cave dwellers. The commandment thus is that we do not delude ourselves into thinking that we can control God by making an image of Him.

3. "Thou shalt not take the name of the Lord thy God in vain."

• It's not that important, we say.

• We use variants: Gee whiz; for crying out loud; by gosh, etc.

4. "Honor the Sabbath Day, to keep it holy."

• Which is the Sabbath Day - Saturday or Sunday?

• Surely you don't mean the whole day?

• Our competitors don't honor the Sabbath, and scheme how to take our businesses or our jobs.

5. "Honor thy father and thy mother."

• Fathers are made out by anonymous writers in the ever more intrusive media to be bumbling idiots who have to be informed and educated by their wives and even their 5-year old children.

• Yet, how can a child honor an addicted parent who beats, rapes, or severely injures him/her?

6. "Thou shalt not kill."

• We have mass killings by "Christian" nations and random killings by individuals.

- We condone and justify murder by childhood traumas, provocation, and "temporary" insanity, ignoring lack of personal responsibility.

7. "Thou shalt not commit adultery."

- We are constantly straining against the bonds and commitments of monogamous marriage.

- The media protest it's only a (movie)(play)(story)(etc.) and their right of free speech allows them to offer it even if it does whittle away or even attack this Commandment.

   Comment: The fact that they are also profiting by such attacks is not considered relevant.

   Comment: They choose to be blind to the truth that man-made laws are only a supplement to the moral laws and practices every society has recognized over the ages.

8. "Thou shalt not steal."

- Unless you can get away with it: everyone's doing it.

- It wasn't all that much.

9. "Thou shalt not bear false witness against thy neighbor."

- Who is my neighbor?

- I need to do this for my own gain, or to cover up my trespass.

10. "Thou shalt not covet . . . anything that is thy neighbor's."

- Coveting is at the root of 8, 9, 7, and 6.

Today, much of the picture I constructed has been torn apart. The prevailing philosophy has changed to what can we get away with. The morality peddled from the TV, the movies, the radio, and books and modeled by some of our prominent people and leaders is now the total antithesis of the Ten Commandments:

I.   Worship as many gods as you please, except the God of the Bible.

II.   Make all the graven images (charms, rabbits' feet, etc.) you
          wish, and believe in their power to help and protect you.
III.   Cheapen and mock the name of God.
IV.   Ignore the Sabbath.
V.   Dishonor your father and mother.
VI.   Kill anyone and in any way you care to.
VII.   Tolerate, condone, excuse, encourage, and commit adultery.
VIII.   Cheat and steal freely.
IX.   Lie as is profitable, even if it destroys others.
X.   Covet anything that is thy neighbor's, and scheme to get it.

All in the name of personal freedom, my "rights", and "You do your own thing". This is a seductive fraud because it suggests unlimited freedom is possible for the individual. Something in us craves freedom and rejects constraints. But total freedom or absence of restraints is not possible even in a primitive society.

It has been tried before. Hear Moses' reminder:

Deuteronomy 12:8 "Ye shall not do after all the things that we do here this day, every man whatsoever is right in his own eyes."

And Paul's counsel to Timothy:

I Timothy 1:8-9 ". . .we know that the law is good, if a man use it lawfully; knowing this. . .the law is made for the lawless and disobedient. . . ."

Do you like acronyms? This one is YDYOT. Pronounce the Y like the Y in MYTH. Pronounce it like IDIOT. We need to be conscious of the limits of this philosophy. On a statistical scale, at one end we have absolutely no external, humanly imposed limits on our behavior. This is anarchy - literally, no government. Its benefit: zero monetary cost. Its drawback: it works only for perfect people, people who always exercise the appropriate degree of restraint in their behavior. Do we want to be like the grains of sand on a beach? Each is free. Free to be swept out to sea and sink to the bottom in total helpless anonymity.

The other end may be totalitarianism, or the limits may be imposed genetically, as in the ants and the bees. Absolute personal freedom in a group conflicts with considerations like how much do we want to be free from the freedoms (attacks and encroachments) of others? I believe the reality is that to have any degree of personal freedom, we must surrender some of our "rights", some of our "freedom".

The society we have inherited has not collapsed yet, but I believe the declines in morality threaten it. Previously, I have refrained from voicing objections because I did not believe I had the right to attack something unless I had adequate alternatives. But now I see others aggressively working to destroy the foundations of our society <u>without</u> proof that the changes they propose are for the better. They deny responsibility for their actions, insist they don't have to prove their actions are not harmful, and it is the objectors who must prove that they are. Surely some of what we have inherited from the agonies of the past is good, useful, and worth preserving. Let me offer a physical example.

- Say you lived in a highrise apartment building and one day discovered a work crew in the basement demolishing the main pillars supporting the building. Would it not be evident that this required at least investigation if not a demand that it cease?

I had thought that it would not be necessary to speak up, that the situation would correct itself, that the pendulum would swing back. But it is still swinging in a direction I can only call negative. The pattern is mockery, debasement and abandonment of the values that made for a cohesive, productive, free society.

And freedom does not simply mean not being a slave or not being in jail. It means being able to walk down the street or along a woodland trail without being murdered. It means not having children, even adults, kidnapped in broad daylight and found dead later because someone else wants to be free to do his own thing (read: robbery, rape, or even just wanton killing).

On a less dramatic, but no less serious, level, freedoms and protections that have been won at enormous human cost are being whittled away. "Human cost" means generations of slavery over thousands of years, cruelty, and death. Perhaps many of us do not feel particularly religious, whatever that involves. But we must become very conscious of how much our way of life comes from most people believing in and practicing the same values, "religious" or not. Vociferous cries say we cannot force our values on others.

- Such criers refuse to recognize they are forcing others to accept <u>their</u> values (or non-values).

- Yet we are already doing this: e.g., we all drive on the right side of the street. But that is not a value, someone objects: it is a practice. Look deeper. If I firmly believe that it is my right at any time to drive on whichever side of the street I feel like, then it is a social value that I agree to relinquish this "right".

We also hear: "There are no absolutes." This assertion often comes up in debates about good and evil. An absolute is defined as something whose existence and nature are independent of any conditions external to it. The assertion is wrong on several counts.

- The statement is itself an absolute. Therefore, it is false, and is its own refutation. So there is at least one absolute. This is an illustration of a simplistic generalization. "Simplistic" means simplified to the point of being wrong.

- The assertion is incorrectly stated. It should say there are no <u>man-made</u> absolutes in the material world. Today's theory in Science can be completely replaced by tomorrow's.

- The material world has a multitude of absolutes: e.g., the Law of Gravity and the analogous Coulomb's Law. No one has found a way of repealing them. We can't even shield ourselves from gravity. Further, without air and water we die. Submerged in water we die. All life forms on Earth die.

- It is improperly applied. Even if it were true in the material world, it is not automatically valid in the nonmaterial worlds of the mind and spirit. The human experiment has been conducted over many more millennia than those of Science so its findings in human relationships have a great deal more statistical validity.

- It is not applied uniformly. Proponents of an absolute right of free speech ferociously attack any efforts to limit speech to that which is appropriate for the occasion. They preach toleration for their views but are intolerant of any contrary views. "Tolerate" has been grossly abused by the wordsmiths. It means to allow without prohibition or hindrance. It is now used to imply that anything and everything should be allowed. For our personal survival, is it not unquestionable that we must be intolerant of murder? Similarly for theft, adultery, and deceit.

Overall, the attack on historical absolutes appears consistent with a deliberate and determined effort to disassemble our social structures and reassemble them in a radically different mode. This is in some cases disguised under the euphemism "transform". We therefore need to be watchful as to what path we travel.

Chapter 6. Explanations for Our Condition

Questions

The written and the unwritten records show it took at least thousands of years to get into our present condition. How did the process start? What controls it?

Many "explanations" have been offered, ranging from total denial of the existence of gods to total acceptance of one God. This apparent progression is certainly not chronological. Some "primitive" tribes have always believed in one god. "Civilized" tribes have often believed in multiple gods.

Why should we be interested in these explanations?

- To remind us of our craving for answers.
- To identify any untried explanation.
- To avoid repeating the failures of the past.
- To become aware of trends.

Explanations

We humans have put together, believed in, and acted upon just about every conceivable variation as to the nature and the origin of the Universe in which we find ourselves. These beliefs are in essence religions. Many have always been willing to believe that some of us have learned Magic, i.e., certain incantations and procedures, that let us control supernatural forces to accomplish our desires.

A. Magic

Frazer (**TGB**, Ch. III, IV) discusses two kinds of magic and their basic errors.

- Homeopathic Magic is based on the Law of Similarity. An image of a person is believed to be that person. Whatever is done to the image will be felt by that person. Its error: assuming that things that resemble each other are in fact the same.

- Contagious Magic is based on the Law of Contact. If one uses another's hair, nail clippings, the navel string etc. to make an image of that person, he can work his will on that person. Its error is assuming that things which have been once in contact are always in contact.

Frazer calls these beliefs forms of Sympathetic Magic, that some secret sympathy allows action at a distance.

> **Comment:** These are stages in the development of logical thinking because they involve noticing apparent cause and effect relationships. Science struggles with the same question of action at a distance by gravity and electromagnetic fields. If properly applied, the Laws lead to Science. They are improperly applied in Magic.

> **Comment:** The more serious aspect is our willingness, ignorant or educated, to believe unreasonable things without experimental proof. Consider the widespread interest in astrology, Ouija boards, Tarot cards, and cults. Critics of both Religion and Science say these ask us to do the same.

Both Magic and Science are based on faith in the order and uniformity of Nature, whose course is determined by immutable laws operating mechanically, and not by the passions or whims of personal beings. Neither appeals to a deity. Both must conform to the rules of their art.

> **Comment:** This is strange. My understanding was that practitioners of Magic believe they are appealing to entities that are at least spirits.

Religion assumes the course of nature is somewhat variable and that we can persuade the mighty being(s) who control it to modify the course of events for our gain.

> **Comment:** But then what happens to predestination?

> **Answer:** We were predestined to make such appeals.

B. Religion

1. Definitions

Frazer defines religion as a conciliation of powers superior to man which are believed to control the course of nature and of human life. Two elements - a belief in higher powers and the use of certain practices - must be present. The former alone is a theology. If one acts from fear or love of God, he is religious; if from fear or love of man, he is merely moral/immoral according to general standards of behavior. A set of practices alone is not a religion.

Dictionaries say a religion is a set of beliefs about the cause, nature, and purpose of the Universe, especially when considered as the creation of a superhuman (supernatural?) agency or agencies (gods) with powers far greater than ours. It usually has devotional and ritual observances and often a moral code for the conduct of human affairs.

> **Comment:** This seems to agree with Frazer. But the "especially" implies that <u>any</u> set of beliefs about the Universe can amount to a religion. Science can be seen as a special case of religion in which as yet no supernatural agency has been acknowledged, although "Nature" is capitalized. I believe the ultimate justification for a belief is that it is "reasonable" and productive: it helps one through life's problems.

> **Comment:** "Religion" incidentally comes from the Latin re + ligere and means to bind again or bind back. It is not clear how this word came to be applied to sets of beliefs, but it does suggest some interesting thoughts. If it means bind again, were we ever bound once and then freed? Or does it mean being tied back to earlier beliefs?

> In either case, there is a definite connotation of being firmly bound to a set of beliefs that hopefully maximize our survival potential in this world and the next. The inescapable consequence is the loss of some of our freedom.

Frazer concludes that humans have gone from Magic to Religion to Science. He warns us (**TGB**, p 712):

> "The history of thought should warn us against concluding that because the scientific theory of the world is the best that has yet been formulated, it is necessarily complete and final. . .the generalizations of science. . .are merely hypotheses devised to explain that ever shifting phantasmagoria of thought which we dignify with the high sounding names of the world. . .science itself may be superseded by some more perfect hypothesis. . . ."

> **Comment:** We might suppose we started with Magic, added elements to get to Religion, and deleted some to branch off to Science. It is by no means certain that Religion is a dead end and that all progress henceforth must come from the path Science takes.

In Religion we must usually deal with an all-powerful God (or gods) who is (are) unseen and whose origins and purposes are mysteries. How can we know if such beings are not deliberately misleading us? A few people (prophets) report receiving thoughts and messages from God regarding human behavior.

Question: Why don't we have prophets today? Perhaps we do. In Science. People like Rutherford, Bohr, Maxwell, Planck, Dirac, and Einstein. Are they false or true prophets?

Question: Why does God speak in whispers? Perhaps He wants to intervene no further than to nudge us to see if we will voluntarily decide to love each other as ourselves and control our free wills.

2.   Secular explanations

Humans have come up with two major classes of explanations: the religious and the secular. A subset of the secular is Science, which explains events on the basis of measurable regularities. These are the impersonal gods (controllers) called "Laws of Nature". There may be forces that are not among those in our Universe we normally see and can understand and that are thus regarded as supernatural but not personal.

3.   Creationist explanations

Many conceptions of gods have been known. Looking over the history of man's religions, we find each contains elements that are in other religions. Millions of words have been written about most religions. Who reads them today? Who remembers the messages of last year's sermons? What problems have been solved by them? Who acts on them? We ask these questions because our review seeks to avoid the Type A error that may result if we discard the past.

The Creationist explanation rests on the authority of the Bible. It states that the Universe was created by a single God, who always existed, and has manifested Himself to humans in three forms. He created an original man and an original woman who disobeyed Him. In accord with His plan, instead of exterminating them, He continued their existence, and subjected us, their descendants, to all manner of severe and horrible tests of our obedience, holding us ultimately accountable for our responses. However, this explanation does not consider some important issues:

-   How <u>did</u> God come into existence?

-   <u>Why</u> did He create the Universe?

- Why does He not go directly to the situation in **Revelation** after Satan is finally subjected to eternal torment in a lake of fire?

- God created Satan; why does He not simply destroy (uncreate) him? Or is He bound by some Law which says once a deity is created it cannot be destroyed?

There are two camps of Creationists. One holds that the Bible is literally true. In particular, the Earth and life were created in six 24-hour days some 6000 to 10000 years ago. The other puts it all into the context of a 4.5 billion year old Earth. Both camps try to use the "scientific method" to bolster and ultimately prove their case. Some of the findings of Science actually are in accord with the Bible.

However, if God has the powers ascribed to Him, all of this intellectual effort is useless. One simply believes that God can do anything and does whatever He pleases. There is nothing to explain. How do we know our poking around in His Universe in our efforts to understand it is not interfering with His plan(s)? Or does His plan include our poking around?

But some of us are not content with this passive position of surrender. Even the Creationists are driven to ask How and Why. So we gradually build a body of conclusions and hypotheses about the Universe and its "Laws" of regular behavior.

C. Science

Science is a systematic knowledge of the material world. Religion can also claim to have systematic knowledge. Knowledge in Science is obtained by the scientific method. This involves identifying a question, formulating a hypothesis to answer it, gathering data by tests and observations, and determining if the hypothesis fits the data. But it also involves having others repeat the experiment and come up with similar data and the same interpretation. Thus, it includes only knowledge based on information that can be detected and/or measured by appropriate sensors.

Question: But is this the only kind of knowledge in our Universe? In the belief system of Science, our Universe was not created, it always existed, it obeys several observable "Natural" Laws, and ultimately everything can be explained by these Laws. Life itself is asserted to have come about through the operation of these laws although no one has proposed mechanisms for its origin and development.

Comment: This seems equally plausible to saying God always existed. Until we ask the next question. What made it start changing? We might say it was always changing, but now we have to ask:

- What set the rate of change?
- What set the direction of change?
- Change usually involves force or energy. Where did these come from?

Recent theories, data, and interpretations are converging on the notion that at some time, 10-20 billion Earth years ago, there was an enormous explosion (The Big Bang) that brought our Universe into being and everything thereafter came about through the operation of the Natural Laws of that Universe. Thus, both Science and Religion now agree that our Universe did not exist forever. Science does not account for some important items:

- What caused the Big Bang?
- What caused the cause of the Big Bang, etc.?
- What was the Universe like prior to the Big Bang?
- If it had a nature, where did it come from?
- What made the laws of our Universe what they are?

Science often has to wait for "insight", which is the direct understanding of something without any reasoning process, to be able to enter into previously unknown areas of knowledge. Thus, we don't invent the Laws of Nature. We merely uncover them. And we cannot change or repeal them. Science also can't prove that its "explanations" have taken into account every relevant fact.

Question: In either case, we have a serious logical problem. How do we decide which explanation to believe in?

Chapter 7. Creationism and Science

Let us now take a preliminary look at two explanations for the origin of our Universe and our Earth.

<u>Creationism</u>

Bypassing the question of whether God is thereby limited in any way, He made our Universe subject to certain laws, some of which we have discovered. One of the arguments for the existence of God is the very order we see in our Universe. Creationists are willing to assume those laws have not changed over 10,000 years. We are also reminded of the source of Biblical truth:

> **II Timothy 3:16** "All scripture is given by inspiration of God. . . ."

> **Job 32: 8** "But there is a spirit in man: and the inspiration of the Almighty giveth them understanding."

The Israelites believed that God spoke directly to them, or to their prophets. It is a source of wonder that God chose not to be clearer in His communications, and that He chose to speak to only a few. But even as I write this, an answer comes into my mind: perhaps He <u>does</u> speak to us all, but most of us choose to ignore His voice. A secular view is that certain Israelites had insights about the nature of things that could not be refuted or replaced by superior insights.

Genesis 1 and 2 give two accounts of Creation. Some say **Genesis 1** gives an overview, while details of man's creation are in **Genesis 2**. The two differ somewhat. The following review focusses on the sequence of events as given in **Genesis 1**. Moses is credited as the author of the first five books of the Bible (**John 5:46-47**), presumably by revelation from God. The comments below are from the perspective that the Bible is literally true. The questions are for the skeptic to answer.

Day 1: God created heaven and earth, then light.

> **Comment:** The earth was without form and void, yet there was a face of the deep waters. There was no dry land, so "without form" must mean no islands, mountains, or valleys. "Void" could mean no life forms. The surface then was spherical.

Comment: The nature of the light is not clear: the Sun and the stars were not created until the fourth day. We now know night is due to the rotation of the Earth into its own shadow, cast by the Sun.

Question: From whence came the inspiration (insight) that on the first day only these two acts were (could be?) done? And isn't it interesting that the ancients had the grand notion that as powerful as God is thought to be He did not create the Universe in one instant? Even though He knew what it would look like when finished.

Comment: The length of time prior to Day 1 is unspecified. Secularists deny time existed prior to the Big Bang probably because it appears impossible for Science ever to describe events prior to the BB.

Comment: "Day" could be a 24-hour day, or a period of time. One needs to be a student of ancient languages (Hebrew, Greek, and Aramaic) and of ancient meanings of words. Note:

> **II Peter 3: 8** ". . .one day is with the Lord as a 1000 years, and a thousand years as one day."

The Bible uses parables and metaphors frequently. I prefer to read it metaphorically as a period of time because this permits us to reconcile the statement with what we observe in our world.

Day 2: God created Heaven, dividing the waters from the waters.

Comment: Confusion enters, for Heaven can mean the place where God, the angels, and the spirits of the righteous after death are found. Here it may mean only the atmospheric gases surrounding the Earth.

The waters below are probably the oceans. But what are the waters above Heaven? Perhaps the ancients did not know of the hydrologic cycle, and thought rain came from a reservoir above.

> **Genesis 7:11** ". . .and the windows of heaven were opened." (The Flood).

Comment: Creation Scientists postulate that the primordial Earth had a dense vapor canopy. Hence all parts of it had a warmer temperature than today. We don't have one now, so the obvious questions are what process supported such a canopy and what made it fail? They say the collapse of the canopy led to the Great Flood. We will examine the Flood mathematically in Chapter 15. However, by the time of

61

Ecclesiastes (c. 200 B.C., or Solomon 930 B.C.) there appears to be some understanding of the hydrological cycle.

> **Ecclesiastes 1: 7** "All the rivers run into the sea; . . .unto the place from whence the rivers come, thither they return again."

Day 3: God created dry land, then plants with seeds.

> **Question:** How did the ancients ever come up with the idea that at one time there was no dry land? If all the dry land were moved into the oceans, what would the elevation of the land be? A half-mile <u>below</u> sea level.

> **Comment:** The plants are growing without sunlight. Were they different from the plants we know today? If so, when and how did they change? The Bible says nothing about a change in their nature. But if we are open-minded, let us consider:

> **Revelation 21:23** "And the city had no need of the sun, neither of the moon, to shine in it; for the glory of God did lighten it, and the Lamb is the light thereof."

> **Comment:** Did the ancients know about the mechanism of sexual reproduction in plants? They spoke of human sperm as seed, but this is inaccurate since plant seed would correspond to the fertilized egg. God thus introduced sex into Earth's plants before He made man. This agrees with paleontology, which finds sexed plants before man.

Day 4: God created the sun, moon and stars.

> **Comment:** With a Sun, we can talk about night and day. What marked off day and night in the first three "days"?

> **Comment:** The ancients did not know about Uranus, Neptune, and Pluto or about the moons of the outer planets. The two moons of Mars were not discovered until 1877. Four of Jupiter's sixteen satellites were first seen by Galileo with the aid of a telescope. None of these are mentioned in the Bible.

Day 5: God created sea creatures and winged fowl.

> **Comment:** The language is:

Genesis 1:20 "And God said, Let the waters bring forth abundantly the moving creature that hath life, and the fowl that may fly above the earth. . . ."

This does not say that God put aquatic life into the oceans. Instead He caused events in the oceans that resulted in such life. While the birds came from the oceans, this doesn't say they evolved from fish, but that there was some process operating that gave rise to different forms of life.

Question: What made the ancients believe life came from the sea?

Day 6: God caused the land to bring forth land creatures after their kind. Then He created man, male and female. He gave all of the animals every green herb for meat, and to the humans every herb and tree bearing seed for meat.

Comment: Here in **Genesis 1:24** land creatures are brought forth from the earth, not the sea. So "moving creatures" must mean only those in the sea, and the land animals had different causes. The ancients saw no evidence of evolution of one species into another. It is specifically ruled out: once created, a species can only copy itself.

Comment: While some of the dinosaurs were vegetarians, others were carnivores. We have a variety of carnivores today that, unlike ruminants like cows, cannot digest grass. As far as we know, carnivores eat other animals and the fish of the sea eat the smaller fish.

Day 7: God is supposed to have rested on Day 7.

Comment: Some say this means we are still in Day 7. This is a narrow reading. The meaning is that Creation was achieved in six periods of time. Whatever God did in the eighth and subsequent periods of His time is not allocated to specific "days".

However, the wording implies God did no more creating on Earth after the sixth day.

There is further information in Chapter 2.

Genesis 2:5 ". . .for the LORD God had not caused it to rain upon the earth. . . ."

Question: How did the ancients conceive of a time prior to which there was no rain? If the Earth was pulled out of the Sun, as some believe, it would have been molten and rain would have been impossible. On cooling, this might have led to the dense atmosphere of water vapor of the Creationists.

Genesis 2: 7 ". . .God formed man of the dust of the ground".

Comment: God did not say, "Let there be man" as He did with all other life. How did the ancients gain this insight if not by revelation?

Genesis 2:18 "And the LORD God said it is not good that the man should be alone; I will make him an help meet for him."

Genesis 2:22 ". . . (the LORD) made he a woman, and brought her to the man."

Comment: If an "help meet" is just a helper for Adam's toil in Eden, why did God not make another man? Why did He make a female, complete with all the parts needed for reproduction? Surely He knew they would discover they were sexed. Especially since they could see the animals carrying out His command to be fruitful and multiply.

Comment: God is omniscient: He knows every detail of future, present and past. Yet He has a plan. We inferior humans know we should draw detailed plans before starting to erect a building. So God is pictured in <u>our</u> image.

God had to know before Creation exactly what everything was going to be like. Since there are trillions of details to be concerned with in the rest of the Universe, those concerning our Solar System would surely be no problem for God to remember. And yet there is a kind of ad hoc character about the creation of Eve.

Genesis 2:19 ". . .out of the ground the LORD God formed every beast of the field, and every fowl of the air. . . ."

Comment: This was <u>after</u> making man, in contrast to **Genesis 1**. Also, the fowl now came from the ground, not the sea. Since He did not explicitly breathe into them the breath of life (whereby man became a living soul), this implies man came into being by a process different from that for all other life.

Science has not yet explained the cause of the differences between man and the rest of life. Its experiments do show differences in intelligence and learning capacity between man and apes. Studies show DNA differences of only 1-2%. Since there is little quantitative difference, what is the nature of the qualitative difference that keeps apes from competing with us in the market place? That is, why didn't apes evolve to levels comparable to us?

Genesis 2:19 "And. . .the LORD God. . .brought them (the beasts and the fowls) unto Adam to see what he would call them. . . ."

Comment: This also is inconsistent with God's omniscience. It says God did not know in advance what Adam would call them.

Science

Science attempts to explain origins by Natural Laws and processes without invoking a supernatural agency. Many believe that Evolution is one of those natural processes. Let's see how Science conforms to the Bible.

Day 1. Currently, the most widely accepted hypothesis is that the Universe resulted from a Big Bang some 20 billion years ago. Also, most of matter in our Universe appears to be hydrogen gas and the astrophysicists believe it takes millions of years to form stars and galaxies from this gas.

The Genesis "light" could be interpreted as radiation over a wide range of the electromagnetic spectrum from the BB. If we consider **Genesis 1:1** as a kind of summary statement, it is then possible to read light as the first act of Creation:

Genesis 1:3 "And God said, Let there be light: and there was light."

Rummaging around in the BB hypothesis, we find an amazing statement. Immediately after the explosion, the Universe was filled with an enormously hot plasma made of various exotic particles. We will review the time sequence later, but for now let's just consider the statement that after about a million years (another article says 500,000 years), protons and electrons <u>suddenly</u> (my emphasis) combined to make hydrogen atoms. No explanation as to why. They just did. It was their "nature".

This process is called <u>re</u>combination, why I don't know because as far as the story goes they were not combined to begin with. Hydrogen absorbs radiation at particular wavelengths rather than continuously across the spectrum. The Universe <u>at that instant</u> <u>became transparent</u>, permitting light to be transmitted. Exactly as it says in the Bible.

Question: How did the ancients conceive of the idea of a sudden appearance of transparency to let light (radiation) through? Of course, it raises the further question of what kind of God was floating around in darkness for who knows how long? Perhaps He doesn't need our kind of light. As noted above, He is His own light.

Comment: I see no reason why it had to occur suddenly in a secular universe, unless "suddenly" means thousands of years. Given a distribution of velocities, the slower moving (and hence less energetic) particles would combine first, and the process could occur over a long period of time as the plasma/gas mixture re-equilibrated.

If there is a connection between these concepts, it is disturbing to think that a day to God might be a million years, not just a thousand.

Comment: The statement in the Bible is: God created. Out of what? There is no attempt at an answer. It has been read as from nothing. **BHT** tells us that matter can be created from energy as a particle/antiparticle pair. Dirac had predicted this earlier. Since the matter in the Universe is positive energy, and gravity is in a sense negative energy, the total energy of the Universe is zero. Nothing, that is.

Question: Which sense? Does this refer to the <u>convention</u> that the potential energy of a particle at an infinite distance from a mass or a charge is arbitrarily taken as zero so that as it falls toward this mass or charge its potential energy is increasingly negative?

Day 2. Eventually the Earth had an atmosphere of permanent gases such as nitrogen and oxygen. If the Earth initially was hot, most of the water would have been present as vapor. As the Earth cooled, the vapor would condense to liquid water.

Day 3. Dry land could emerge by various processes:

- Geological processes could distribute solid matter nonuniformly;

- Much water could be stored as ice or tied up chemically; and

- Water levels would drop as water was gradually lost to outer space.

Day 4. This simply records the condensation of gaseous clouds into organized galaxies and solar systems. It is out of sequence with the view of Science that stars had to come before planets. However, astronomers believe that stars are continually being born all over our Universe.

Day 5. This agrees with the notion that life began in the sea, in which various chemicals could contact each other and be acted upon by thermal energy and radiation. It is not necessary that this occurred over the entire Earth. It is sufficient that local conditions were favorable.

Day 6. The ancients said land creatures came after sea creatures, as do the evolutionists. The ancients also believed that man came last.

> **Comment:** "Dust" implies mineral dust, perhaps silicates or phosphates. The ancients burned enough people to know that ashes were left. Our bodies do not use silicon or aluminum. They do use phosphates. It seems the ancients were way ahead of us.

> A recent speculation is that life began with the aid of clays, as stated in the Bible. Various aluminosilicates are active catalysts for chemical reactions, though at high temperatures. The scientists are far from explaining how life began if it was not created or what caused such an incredible proliferation of life forms.

Is there anything really strange about all this? Even very young children are driven to ask questions about the world around them. Reflect on the countless generations of humans observing their world, noticing sequences and patterns. When they had accepted the fact that sexual union of parents is what begot children, and that one pair of parents could have many children, they had to infer that the population was smaller in earlier times. An obvious question: was there an original pair of parents?

> **Question:** Did this original pair come already stocked with several hundred million different ova and several hundred billion different sperm?

> **Comment:** Rather than postulating an original pair of parents for every species, which then involves Creation, Science has to postulate an original life form that evolved by still unknown mechanisms into all the others. This postulate still requires some recurrent events to cause mutations.

Then they asked a more difficult question: which came first - the sheep or the grass. Shepherds see their sheep eating grass; the grass doesn't eat the sheep. So the grass, i.e., plant life, must have come before land animals.

Charles H. Peterson

Comment: Who are we to say how God does His work? Just as we set fires in boilers to generate steam to make electricity to run motors and appliances without having to attend to each one constantly, so God could have put the Sun in as a gigantic steam boiler evaporating water from the oceans that would fall as rain and snow to put essentially pure water on the land. Is God to be more of a slave to His creations than we are to our machines?

Comment: God could have made millions of "Earths" scattered throughout the Universe and we are just one of His acts of creation, perhaps one of his experiments. To what extent are we in God's image? Is it like looking into a mirror and seeing a two dimensional image of one's self? So is man a three dimensional image of a multidimensional entity?

Overall, surprising parallels exist between the Biblical account and the Scientific account. Science might see the Bible as an evolution of man's explanations for our Universe. If the Bible is wrong, why has it survived? Think how many other religions are extinct. . . .

Science does appear to be somewhat arrogant in denying the possibility of truth in the beliefs and explanations of non-scientists that are not readily susceptible to verification by the methods of Science.

- "Old wives" remedies: herbal remedies
- Native cures and methods: acupuncture
- Biblical explanations
- Miraculous healings

Creationists seem arrogant in distorting, denying and ignoring information about our Universe, particularly as to its age. Fundamentalists insist on the literal truth of the Bible despite several things that do not appear defensible and despite textual inconsistencies.

God could have done just two acts. First He created the Universe, perhaps in a BB, along with its Natural Laws. Then He inoculated it with life.

Why should we conceive of an infinite God who is yet a slave so that He has to tediously micromanage each of His millions of creations? What He has done with respect to living things is more elegant: He created a basic formula through DNA that works out its destiny through the laws He also gave. This would be consistent with believing that God rested on the seventh day.

Speculation: Perhaps He is interested in determining the minimum specifications for 3-dimensional material beings capable of relating to one another. Perhaps Christ's appearance on Earth was a midcourse correction, another chance for us, before He decides to terminate the whole experiment of Man.

Chapter 8. Nature of Our Solar System

So much for a quick look at some of the teachings of the Bible and of Science. We must dig deeper into them before we can choose which to follow, if either. In this book, we will not treat theology in depth. In the past we were forced to believe, on pain of torture and death for heresy against the Christian religion, that the Earth was the center of the Universe on the authority of the Bible, and everything we saw on looking up into the sky - Sun, Moon, planets and stars - rotated around the Earth.

While most of us now believe these ideas are false - I believe the Flat Earth Society has disbanded - we are not familiar with the newer concepts, except for what is in popularized writings. No one of these gives enough information for the reader either to evaluate their content or construct a comprehensive, coherent, convincing picture of our Universe.

We need to review the evidence for what we are told about our Universe. Both in Science and in Religion, one idea builds on another, so if we cannot accept the conclusions we can go over the chain of ideas to find the weak link. It is more difficult to apply this process to Religion, because it rests on faith in a God.

Science has captured the public imagination with its achievements in utilizing the forces of Nature. It exerts a powerful influence on what we do, how we use our resources, and what we believe. Listen then to:

Matthew 5:41: "And whosoever shall compel thee to go a mile, go with him twain."

Science promises us a much more complete, consistent and credible set of answers for our questions. It also offers measurements which in theory anyone can repeat for themselves. Let's see what Science has to offer to explain our condition, what path it wishes to lead us on, what hope it offers to ease our pain and suffering. If we find and refute a weak link, we may have to believe in some kind of supernaturalism.

There are several problems. The literature is voluminous, and I have not found an integrated picture. Each author writes about a part of the picture, and the descriptions are changing. Statements and figures at times disagree, even within the same text. Too many details are omitted, so the non-expert cannot possibly evaluate the truths of the assertions made. Further, the assumptions made are rarely mentioned later as limitations on the conclusions.

My understanding of some of these matters is no longer primitive, but high energy particle physics is largely beyond me. Nevertheless, looking at a problem from a different point of view can be useful. We certainly can ask what are we sure we know and how do we know it? What can we say about our world that we can verify?

Since the Bible of Science is constantly changing, what I offer here is an initial attempt to construct an integrated picture without becoming a theoretical physicist or any other kind of expert. I raise many questions. Perhaps they can be resolved by continued study with help from others who also study them.

The information in this and the next few chapters may seem unrelated to our question. It may also seem highly technical. It does deal with technical matters, but the mathematics is mostly simple arithmetic. We will look at some of the measurements of Science, with an emphasis on concepts and "explanations". They present Science's picture of our physical world.

Generally, you will be able to focus on the text and skip over the mathematics presented in the indented sections in the text if you are so inclined. Longer treatments are included in Exhibits at the end of the various chapters. What we will do is take a walk through this world to see what kind of a philosophy, if not also a religion, we are offered. Religion rests on God. What does Science rest on? Perhaps you will find that you too can ask many questions. . . .

Shape of the Earth

A. Flat Earth

Let's start with our immediate world. What do we see? In any flat land in the world, we could look to the horizon and see the sky meet the earth. Looking up, we cannot see an end to the sky. Put some clouds in the sky. Watch them float off to the horizon and we see them touching down there. So it would seem we live on a flat Earth under a hemispherical sky. That is common sense.

This is close to the view of Thales (c640-546? B.C.): the world is a hemisphere resting on an endless expanse of water and the Earth is a flat disk floating on the flat side of the interior of this hemisphere. He did not say what was holding the water in place, or worry that a slight disturbance in that infinite expanse of water could obliterate the works of man. Somehow the ancients also never talked about finding the spot where the clouds did touch land.

This early view had elements of truth. We now believe the "solid" Earth we stand on consists of not one but several plates about 30 miles thick floating on a molten core. They are called tectonic plates.

> **Comment:** "Tectonic" means "pertaining to the structure of the Earth's crust". So what we stand on is made of crustal crusts. It's kind of pretentious, like the weather reports about rain "activity".

> Then we find "tectonic" comes from the Greek "tekton", meaning carpenter! So it really means plates made by the Carpenter. Religion, and in particular Christianity, is sneaking in the back door of Science. God must get a chuckle out of this.

B. Round Earth

Anaximander (c611-547 B.C.) taught that the Earth hung freely in the center of the Universe supported by nothing but held in place because it was equidistant from everything (**TSC**). But he also taught that the Earth was a cylinder and the Sun, Moon and stars moved in circles around the Earth.

> **Comment:** Here is a glimmer of the Newtonian idea that bodies at a distance can influence each other. Also take a moment to note:

> **Job 26:7** "(He) . . .hangeth the earth upon nothing."

> How did the ancients come to such an astounding conclusion? All their experience would argue for having the Earth supported upon something material. How many of us have thought of this?

The writer of Job is thought to be an Israelite some time between Solomon and the Exile (974-586 B.C.) but the material may date back to before 1000 BC. Job is thought to be the oldest <u>written</u> book of the Bible.

This says nothing about when the ideas were first expressed orally. We know lengthy accounts, histories and stories can be memorized and passed down through the generations.

We know our world is three dimensional. What shape is the Sun? We see it as a circular disc. It must have some thickness. Is it a cylinder seen end on, perhaps as thin as a coin? From every place on Earth. But it is more reasonable to assume the Sun is a sphere. And the same goes for the Moon. What about the Earth?

Eclipses give us some information. Anaxagoras (500?-428 B.C.) taught that the Moon is eclipsed by the interposition of the Earth (**TSC**). (He was condemned to death by the Athenians for teaching that the Sun was not a god but a red hot mass of stone, so he fled.) In an eclipse of the Moon, we see the leading edge of the darkness advancing over the face of the Moon is curved. So the Earth is at least a circular plate. Is it a sphere? One way to tell is to sail around it. Or, if we drive toward a city we can see the tops of buildings in that city before we see their bases. Or, we see the tops of mountains before we see their bases. We have teachings from other Greeks (dates are B.C.).

- Philolaus (b480): Earth and the planets revolve around the Sun.

- Parmenides (c450): The Earth is round.

- Aristarchus (c310-230): The Earth and the planets revolve around the Sun in circles, while the Sun and the stars are stationary.

Aristarchus concluded that the stars must be at least several million miles away from the Earth because he could not detect any parallax in any of them.

> **Comment:** A simple example of parallax occurs when we close one eye, extend our arm and by sighting from our other eye line up a finger and some object in the distance. Closing this eye and opening the other, we see that the object has apparently moved.

We are led to conclude the Earth is round, even though we don't understand why it does not fall down. What is "down" if the Earth is round? There is also the objection that people on the other side of the Earth are standing upside down, which is "obviously" ridiculous. If true, it means we are standing upside down relative to their point of view. Common sense was misleading. We are forced to ignore our own observation and accept something we can't understand.

This exercise illustrates the process of discovery. We can learn some things through observation, and by daring to think the different, the "impossible". But we can move on to further understandings only by postulating certain other "facts" and relations. We must, of course, test those postulates and hypotheses and accept only those which are consistent, reproducible, and are open-ended, i.e., have predictive capability.

Size of the Earth

TSC notes that Dicaearchus of Messana (somehow) concluded the circumference of the Earth was (the equivalent of) 30,000 miles. He lived about 320 B.C.

Eratosthenes of ancient Greece (276?-195? B.C.) made an early accurate estimate of the size of the Earth. It is questionable whether any of us could duplicate this feat today. It involves having leisure time, curiosity, and an understanding of some simple geometry involving parallel lines and similar triangles. It also involves having a deep well. Who has seen a well? Today many of us are city-dwellers and have centralized water systems.

The key is simple. Draw two parallel straight horizontal lines AB and CD, with A and C being the left ends. From point E above AB draw another straight line EF crossing AB at G and CD at H. This forms a set of four angles at G and another set at H. Note that the angles that are vertically opposite each other at each intersection point are equal, like AGE and BGH. While this looks reasonable from the diagram, it is easily proven.

> **Proof:** Each angle is supplementary to the same angle, BGE. Two angles are supplementary when they add up to 180°.

Next, the angles whose sides are parallel or identical are also equal, like AGE and CHE. This follows from the axioms about straight lines and parallel lines. These two observations prove that the alternate interior angles, like BGH and CHG, are also equal.

Eratosthenes noticed that at noon on the summer solstice the Sun shone directly down a well in Syene, Egypt, on the surface of the water below. Syene was or was near Aswan in Egypt. Now this at first is simply an observation of something unusual that merely arouses an idle curiosity. However, he believed the Earth was round. Then he realized that the path of these rays of the Sun was exactly lined up with a radius of the Earth!

> **Description:** Draw a circle to represent the Earth. Draw two radii, and extend both up past the circumference of the circle (the surface of the Earth). The rays entering the well then lie along one of the extended radii.

In Alexandria some 500 miles to the north was an obelisk. Either he had someone measure the length of the shadow of the obelisk, also at noon on the same day, or he did it himself the next year. Noon was presumably determined by a sundial which is positioned with respect to the Sun. The obelisk, on the other hand, is positioned relative to Earth's gravity to be perpendicular to the Earth's surface.

Description: Now position the obelisk along the second extended radius. Eratosthenes also believed that the Sun was far enough away from the earth so that the ray past the top of the obelisk would be nearly parallel to the one down the well. Draw a line from the top of the obelisk parallel to the first radius down to the ground. The angle between this line and the obelisk is measured by the ratio of the length of the shadow and the height of the obelisk. The angle was 7.0 -7.5°. It is equal to the angle between the radii.

How did he know how tall the obelisk was? Maybe the builder told him. But he could also determine it for himself. Most of us are familiar with lining things up: building a brick wall; sighting along a gun barrel. We visualize a straight line connecting two points.

Description: Suppose there is a vertical pole at point B near the obelisk. Pick a point P on the pole and stand at some point A so that P appears to be in a straight line of sight to T, the top of the obelisk. Our eye is at a distance AE above ground. Measure the distance AB from A to the pole and also the distance AC from A to the obelisk. Dividing the the length TC minus AE by the length AC gives a measure of the angle between our line of sight and the horizontal. By similar triangles, the same angle is measured by the ratio of (PB minus AE) to AB. Multiplying by AC and adding AE gives us the height of the obelisk.

He knew the distance between the two towns. One source says 5000 stadia (about 575 miles); another says 500 miles. It was probably estimated from the time it took a camel caravan to travel between the two locations. He calculated the Earth's circumference as 24,662 miles vs today's figure of 24,847 (**TSC**). This is curious since no one knows exactly how long a stadium was. From the circumference, he could calculate the diameter of the Earth as 7850 miles. We would get 7909 miles.

Distance of the Earth from the Moon

Aristarchus estimated the diameter of the Moon as 1/3 that of the Earth with an error of only about 8% (**TSC**). He may have done this by comparing the curvature of the shadow on the Moon during a lunar eclipse with that of the Moon's circumference. He already believed the shadow was due to the Earth. The ratio we would get is 2160 miles/7909 miles, or 0.273. So his error was more like 18%. Still rather good. Eratosthenes could calculate the Moon's diameter as 2617 miles.

Sighting at the Moon, we can estimate its distance from us as also about 100 times its diameter. So this tells us the Moon is about 261,700 miles from the Earth. The mean distance is 238,857 miles. Hipparchus (190?-125? B.C.) estimated the Moon-Earth distance as 250,000 miles - an error of only 5% (**TSC**).

To observe the distance/diameter ratio, it is convenient to use a template having a series of circles of different diameters. I found a 5/32" circle at 18" just covers the Moon. This yields a ratio of 115 vs. a true mean value of 238,857/2160 miles, or 111.

Distance of the Earth from the Sun

A total eclipse of the Sun by the Moon gives us some ratios. The diameter of the Sun divided by its distance from the Earth is equal to the diameter of the Moon divided by its distance from the Earth. Clearly, the Sun has to be bigger and farther from the Earth than the Moon.

The best the ancient Greeks could do as to the Earth-Sun distance was 600 Earth diameters, about 4.8 million miles (Hipparchus). By sighting at the Sun, we can estimate the ratio of its distance from us to its diameter. It is convenient to do this near sunset or sunrise when the Sun appears as a dull red disc. A 3/16" circle at a distance of 20" just fits the disc of the Sun, which makes the ratio 107. Today, we use radar and measurements of the Earth's orbital velocity and other methods. Using the accepted value for the Astronomical Unit (mean Earth-Sun distance) in Exhibit 9.1, our estimate of the Sun's diameter would then be 868,700 miles. The accepted value is 865,000 miles.

> **Comment:** It is curious how often the number 100 crops up in the physical dimensions of our Universe. For both the Moon and the Sun, the distance/diameter ratio is about 100. We will see other examples in our review.

Other Information

This may be as far as we can go in understanding our Universe scientifically without the aid of modern technology. The things we have been talking about are things we can see for ourselves. How can we hope to explain things we can't even see?

The conclusions from technology are less obvious but have been developed from much study, observation and experimentation. The next chapters offer a path through the overwhelming mass of technical information accumulated in the last century or so on the nature of matter, energy, our Universe, and our origins. The objectives are:

- To present some conclusions about our Universe.
- To identify assumptions.
- To pose questions to arouse curiosity.
- To indicate how much remains unexplained.

What is the basic message? Science has given names to several phenomena and quantities but cannot explain what they really are. Some of the definitions are arbitrary. Science does not know why these quantities and phenomena exist. There are inconsistencies, e.g., most references say the lowest possible temperature is absolute zero (0 Kelvin), yet negative temperatures are discussed. Science asserts its explanations are superior to those of Religion.

Religion is challenged to prove its claims, which may be impossible. No one in modern times has devised a way to induce or force his God to appear in tangible form or cause some non-natural event at his will. "Proof" is then based on indirect evidence which, nevertheless, may be as persuasive when compared to some "proofs" of Science.

Cosmology and related sciences, on the other hand, are increasingly difficult to challenge, not because they have been proven beyond a doubt, but because their methods involve advanced technology that require much study to understand before a critique is possible. Perhaps the effort to understand presented herein includes some erroneous conceptions and interpretations. The effort is necessary because physicists have yet to formulate their claims in comprehensible terms so that non-physicists can afford to accept their claims. The subject matter is too important for acceptance of those claims on the mere assertions of those making them.

As you read this material, do not be held up by the equations. Details of calculations are included to provide firm support to the conclusions rather than merely making assertions. You can skip over the details and continue with the text following them. However, the time and effort spent in understanding the calculations will add to your sense of conviction.

Chapter 9. Technology: Definitions and Relationships

General

Our objective is to identify the ultimate bases for Science's claims. When we search technology references for answers, we find our questions are not fully answered. Checking other texts, we find they merely repeat the same incomplete "explanation". We may have to go back to the original experiments to understand what was actually done. This is a task for more than one lifetime. To aid in this, we need a working understanding of the basic quantities used by Science and how they are measured. In Exhibit 9.1 are current values for several of them.

Numbers

Most readers do not deal with equations, symbols, and mathematical notation, and hence are unfamiliar with these. Most of what is offered here involves only simple arithmetic. Exponentiation is a special case of multiplication: instead of writing $100 = 10 \times 10$, we write $100 = 10^2$. The 10 is the base; the 2 is an exponent showing the number of times we are to multiply the base by itself. Similarly,

$$0.01 = 1/100 = 1/(10 \times 10) = 1/10^2 = 10^{-2}.$$

The material deals with very large numbers and very small numbers. We need a shorthand way of writing them. Using E to mean "exponent using base 10":

$100 = 10^2 = 1.0\text{E}+02$, means 1 times (10x10),

$0.01 = 10^{-2} = 1.0\text{E}-02$, means 1 divided by (10x10).

Usually, numbers are written with one leading digit followed by a decimal point and the rest of the digits in the number. Often I will depart from this so that the exponential part can be read more easily.

E+03	thousands	E+12	trillions
E+06	millions	E+15	quadrillions
E+09	billions	E+18	quintillions

Precision

Most numbers in Science are not exact integers. They are measured to varying degrees of precision. If we were counting U. S. money, 1.24 means exactly one dollar and 24 cents. In Science, it means 1.24 ± 0.01. If we are rounding off the results of calculations, the 1.24 was anything in the range 1.235 - 1.244.

The result of a calculation is no more precise than the least precise number used. Exhibit 9.1 gives many numbers to a large number of decimal places because that is what is in the literature. In using them, consider these examples:

$$1.04 + 0.0000012345 = 1.04$$
$$1.04 + 0.12345 = 1.16$$
$$1.04 \times 0.0000012345 = 0.00000128388$$
$$= 0.00000128$$

There are five significant digits in the second number in each case, but most are irrelevant since in Science the first number means 1.035 to 1.044. While calculating, one carries one or more extra digits and rounds off the final answer. Thus, in the multiplication example, the result is rounded off to three significant digits. Caution: precision deals only with how exactly we can measure a number, not its correctness.

Arithmetic Operations

Two quantities written next to each other, like ab, means $a \times b$; a/b means a divided by b. Where parentheses are used, all the operations within them are to be done first and the result used for the next operation. Thus, 1+4/5 is not clear. It could mean 5/5, or 1; or it could mean 1+0.8, or 1.8. If we mean the former, we have to write (1+4)/5. The symbol $<$ means what's on the left of it is less than what's on the right of it. The symbol $>$ means greater than.

As another aid to understanding the material, usually the numbers are followed by units to show what the numbers represent. For example, if we travel 30 miles per hour for 2 hours, we have travelled 60 miles:

$$(30 \text{ miles/hr})(2 \text{ hr}) = 60 \text{ miles.}$$

The units are treated like numbers so that the hours cancel out, showing that the product is a distance.

Charles H. Peterson

Some Cautions

Science has gone so far in its search for understanding of how the Universe is constructed and operates that it tends to omit mention that its theories are just that. At <u>any</u> time, some new observation can disprove a previous hypothesis. An assertion like "Protons are made of quarks" really means "according to present hypotheses". Such statements create a bias in the reader's mind that Science always speaks the truth. Therefore, in reading the technical literature one has to be alert to what is merely an assertion or a hypothesis as distinguished from an accepted fact.

Science often uses an important simplification. Because the bodies under study are usually very small compared with their separations from one another, they are treated as mathematical points. This may be suitable for describing the interactions between bodies, but it prevents understanding their structure.

Measurements

Science rests on measurements. It strives to be definite not only about X being bigger than Y but also by how much X is bigger than Y. Its conclusions stem from its ability to measure intervals like differences and changes of distance, time, temperature, etc.

With the concept of time intervals, we can say that B occurred exactly so many units of time after A. This leads to another concept: the time rate of change. We can talk about how fast something changes its position, or how quickly something changes its composition in a chemical reaction. Or even how fast a rate changes.

All measurements involve a unit of measurement and the number of units: (3)(feet); (10)(kilograms); (60)(miles per hour); (1)(firkin per furlong).

The standard of length was once set by arbitrary references, such as the length of the king's arm, or the width of his palm. In recent times, we had a platinum–iridium bar kept at constant temperature and marked with two scratches to indicate a meter. Today Science assumes that the velocity of light, c, is constant. A meter is then defined as the distance light travels in $1/c$ seconds, where c is in meters/second.

Comment: Science has been forced to construct its edifice of knowledge upon - Light. Recall:

Genesis 1:3: "God said, Let there be light . . . "

Light shows a dual nature: waves and particles. Science cannot explain how light can show one nature now and the other later. Science cannot even produce a hypothesis, let alone the true explanation. <u>Both</u> have never been shown in the same experiment. Science is not troubled by this, shrugging it off as a "fact".

Religion has a more difficult task: the Christian God is said to have <u>three</u> distinct natures. There is much resistance to accepting this as a fact, and the concept is even ridiculed. Although the Bible reports God as referring to Himself as "us", there is only one occasion when we are told all three natures were evident at the same time. That was at the baptism of Christ (**Mark 1:11; Matthew 3:17**). **John 1:32** reports only two of the natures being present. We also have Jesus saying:

> **John 16:7** "It is expedient that I go away: for if I go not away, the Comforter will not come unto you; but if I depart, I will send him unto you."

> **Question:** Why did Christ have to leave? Why could not both He and the Comforter be present together?

Tools of Science

The tools of Science are based on electromagnetic radiation. The optical microscope uses visible light. The telescope and the camera extend into the infrared. The spectrograph goes also into the ultraviolet. The electron microscope uses electron beams. Then there are probes (beams of X-rays or particles) and devices like meters, centrifuges, accelerators, and cloud chambers.

Relationships

Science looks for simple relationships ("Laws") that always hold. They are of several types. A quantity X depends on the values of constants, usually represented by letters such as A, B, C, or on other variables such as R, S, T, etc.

<u>Type</u>	<u>General Form</u>
1. Constant	$X = A$
2. Additive	$X = R + S + T + \ldots$

3.	Multiplicative	$X = R \times S \times T \times \ldots$
4.	Power	$X = R \times R \times R \times \ldots = R^n$
5.	Exponential	$X = e^R$

The letter n is the number of R's being multiplied together. The letter e is the base of natural logarithms: 2.71828 . . . Examples are in Exhibit 9.1 at the end of this chapter. We can combine these types.

Comment: These are called Laws of "Nature", whatever that is, or Laws of God, whoever He is. They are unchanging, in contrast to man-made laws.

Let's get something clear: we humans have not invented a single one of these Laws. Nor are we able to cancel any of them. For a long time, we did not even know they existed. They were there before we were. Science does not often humble itself to publicly acknowledge this. It has only described them sufficiently to be able to use them.

Why does Science seek Laws? To relate effects to causes. Why? So that we can predict behavior. Why? Curiosity. Power. Ego satisfaction. Fear of the Unknown. Ultimately, survival.

How are they developed? Ultimately, they are all based on experience, so they are empirical approximations. We don't know why they are valid but they work.

Example: Observations show that the force on a body due to gravity varies directly as the mass of the body. Twice the mass m, twice the force F. Similarly for the gravitational attraction g. So the force F = mg. But force is a completely different quantity than either m or g. It is like asking what do we get if we multiply 3 apples by 5 oranges. We need a proportionality factor k:

F = kmg.

Next we need either a known force to evaluate k, or we choose a value for k so we can calculate F. We do not have a known force, so k is arbitrarily set to 1 newton/ [kilogram-(meter per second^2)].

Other relations may seem to be derived from theories, but they are only combinations of empirical laws. All involve one or more basic assumptions and/or simplifications that work because Science is concerned with only a part of the range of values of the variables.

The philosophic basis for this trial and error process is that Science should be free to change its perceptions of the Laws of Nature to fit new data. Unfortunately, people impulsively apply this idea to human relations without proof of validity, and ignoring the thousands of years of our bloody experiments to find rules of behavior for humans.

Definitions

We begin with definitions, which amount to a cast of characters. My main sources of information for these and the concepts and relations discussed are **WTID**, **EST** and **CVEP**. Letters in boldface are convenient for making references, all of which are cited alphabetically at the end of this book. The dictionary is generally available and its contents usually understandable. Some definitions are from the **RHCD**. The encyclopedia is a convenient source of currently accepted facts and opinions on science subjects.

Especially useful is a physics text. Borowitz and Bornstein (**CVEP**) have done a marvelous job in making college-level information on concepts comprehensible by providing details of derivations and numerical examples. But even here parts are still not clear.

Definitions of Basic Quantities

A. Space

"Space" is a three-dimensional entity that extends without bounds in all directions and is the field of physical objects and events and their order and relationships. It is, alternatively, a mathematical model that pictures physical space as three-dimensional, as partly filled with material bodies, as capable of existence if all physical bodies were destroyed, and as determining, but not determined by, the relative positions and direction of material bodies.

> **Question:** Has this told us what space is? No, because to understand this, we also have to know what the meanings of all the underlined words are.

> **Clarification:** Suppose we want to define "rain". We might say rain is what makes the road wet. But what is rain? We can answer, rain is water. The form of this definition is "X is A", where A is not simply a synonym for X. Saying that "X contains A", or "X causes A", or "X has the property A" helps identify X but does not reveal its ultimate nature.

83

So let's examine this definition more closely. We touched on some of these terms in Chapter 4 in discussing "truth". Now we need working definitions. Notice that efforts to define a word often introduce other words that in turn require definition. Because the trail gets complicated, all the underlined words in the definition of "space" are defined first.

Term	Definition
Entity	Something that has separate and distinct existence and objective or conceptual reality.

Comment: Five more words require definition.

Field	A large unbroken expanse of anything.
Object	Anything that is visible or tangible and is stable in form.
Physical	Relating to material things or natural laws as opposed to things mental, spiritual, or imaginary.

Comment: This seems to be derived from "physics", the science dealing with matter and energy. "Physical space" is certainly not tangible. What is non-physical space? "Space" is sufficient.

Comment: So space is not an object, not being visible or tangible. It is at least an inanimate thing.

Material	Formed of or consisting of matter.
Body	A separate physical entity, mass or quantity as distinguished from other masses or quantities.
Existence	The state of having being, especially as considered independently of human consciousness.

We now pick up the second and third levels of definitions, beginning with "entity".

Something	Some thing.
Thing	1) An entity or object that is not precisely described; 2) A material object without life or consciousness.

Term	Definition

Comment: So space is an entity, which is a thing that is an entity. This series of definitions is circular, clarifying nothing. Or, space is some thing that is not (cannot be?) precisely described. Is this the best we can do?

Comment: This does not rule out ghosts: visible and stable entities. By 1) a ghost is a thing because it is not precisely described but by 2), it is not a thing, being nonmaterial but by some accounts having a kind of consciousness.

Objective	Existing in the <u>sensible</u> world and being verifiable, especially by scientific methods.
Subjective	Existing in the mind of an observer rather than belonging to an object being observed.
Sensible	1) Readily perceptible by the <u>senses</u>; material. 2) Perceptible to the mind.
Sense	A consciousness of a <u>stimulus</u> that may be subjective or objective; a faculty for perceiving stimuli.
Stimulus	Loose translation: Something that causes a response by a sensor.

Comment: Without a stimulus and a mechanism to sense it nothing can be said to exist. Conversely, if the sensors of our bodies or of our machines detect a stimulus, we must by our own definitions accord existence to that stimulus. Experience has shown that our survival is enhanced when we seek the cause of a stimulus (even a hallucination).

Conceptual	Pertaining to an idea of something formed by mentally combining all its characteristics.
Idea	A product of mental activity, usually a serious mental concept.

Comment: So mental activity produces a concept which is an idea which is a serious concept. Circular.

Term	Definition
Reality	Existence of something independent of ideas about it.
Expanse	An uninterrupted space. So "space" is a field which is an expanse of something which is an uninterrupted space. Again, circularity.
Tangible	Perceptible as materially existent by the sense of touch.

Comment: This seems to say a cloud is not an object even though it is composed of droplets of water. It is visible but not tangible or stable in form. Perhaps we have to understand that an object merely has to be reasonably stable. How about a rainbow or a mirage? These are visible and (reasonably) stable, but not tangible.

Matter	1) Something that <u>occupies</u> space.
	2) A <u>physical</u> or <u>corporeal</u> <u>substance</u> especially as distinguished from incorporeal substances, as spirit, mind, qualities, actions, etc.

Comment: Note that matter is just "some thing".

Occupy	To fill up space.

Comment: Here finally is a clue about "space": it can be filled, or it can contain "things".

Question: Does placing objects in space push aside some space? And does that displace other space? These questions are reasonable since Science wishes to give space structure in the form of strings.

Corporeal	Material; tangible
Substance	1) That of which a thing consists.
	2) Physical matter or material.

Comment: So a "physical object" is a material thing that has a stable form, occupies space, and is visible or tangible, which means it is made of material which is made of matter which is a corporeal substance which is a material material.

Term	Definition
Being	The state of having existence or reality.

Comment: So existence is the state of having being which is the state of having existence. This is useless, being circular.

Sorting through the above, it seems we have to start with ourselves as entities that have some powers of perception of the Universe in which we find ourselves.

- We have the sense of touch, which includes a perception of temperature, so we can perceive things outside of our skin.

- We call these things objects because our fingers don't go through them, i.e. they and the objects can't occupy the same space at the same time.

- We believe they are there (exist) whether or not we touch or think about them. We say then that these objects are real, or have reality.

- If all these objects were removed, what would be left is space.

- We also perceive other things through our senses. All of these perceptions depend on the properties of matter or energy. They also have reality, since others can perceive them, and often replicate them.

- Some report other perceptions, but these are outside the scope of this book.

Conclusion: The truth is that Science does not know what space really is and, if it is something more than nothing, where it came from. Let us think of space as a nonmaterial nothing because it has no measurable length, width, breadth, mass, temperature, radiation or absorption. It seems to be continuous: there is nothing to mark where it begins or ends. Also, objects do not displace space.

B. Matter

Science does not know what matter is.

The deeper we probe, the more incomprehensible and elusive it becomes. It composes the observable Universe of physical objects and together with energy is believed to form the basis of all objective phenomena.

> **Comment:** It is thus admitted that matter and energy have not been <u>proven</u> to form the basis of all objective phenomena. This is only a belief, which is one of the components of a religion. A material body is made of tangible matter, which is something perceptible to touch. If I place my finger on a stone I am aware that something is resisting my finger, but I cannot feel anything if I put my finger in a (nonmaterial) light ray. Matter has several general properties, such as:
>
> - Volume, mass
> - Impenetrability, divisibility, porosity
> - Compressibility, expansibility, elasticity
> - Inertia, mobility.

Impenetrability means material bodies cannot <u>occupy</u> the same space at the same time. Matter consists ultimately of elementary <u>particles</u> of relatively few kinds. We elaborate on "occupy".

> - Occupy: Besides filling up space, this seems to imply excluding other bodies. Two light beams can intersect, i.e., occupy the same space at the same time. This shows that light is not matter.
>
> - Particle: A very small portion of matter. An elementary particle is a subatomic unit of mass and energy that has a characteristic mass and energy and quantum properties, such as charge and spin.

Science recognizes three main states of matter: solid, liquid, and gas. Simple distinctions are: a gas has no shape but spontaneously expands to fill its container; a liquid takes the shape of its container but does not tend to fill it; and a solid has its own shape.

C. Inertia

Science does not know why matter has this property.

Inertia is the tendency of bodies at rest or in motion at constant velocity to remain in that condition unless acted upon by outside forces. Considering the bizarre concepts of space currently offered, we can't be sure that inertia is an exclusive property of matter - it may be an interaction with space.

D. Elements, Atoms, and Molecules

An element is one of 103 substances that differ in their properties and from which all known matter is made. An atom is the smallest quantity of an element that still has the properties of that element.   Molecules are combinations of two or more atoms, which may be the same or different.

E. Time

Science does not know what time is, or whether it has a beginning or an end.

The definition offered is that time is the dimension of our physical Universe which orders the sequence of events in a given place.

> **Comment:** This concedes there is some kind of Universe other than the physical. It is also a functional definition, not a fundamental one.

Science says time began with a Big Bang 20 billion years ago. Or 10 billion. Science cleverly beguiles us with assertions as to how events proceeded <u>after</u> the Big Bang, hoping to distract us from the question of what was going on <u>before</u> the Big Bang. Something had to exist <u>before</u> it could explode in a Big Bang.

> **Comment:** In contrast, space is defined as without bound. Since we are told we live in a 4-D "spacetime", why should not time be granted the same freedom from limits? It is inconsistent and thus irrational. Perhaps Science views time anthropomorphically. Time in Religion is unbounded, which appears more consistent, symmetrical and therefore more reasonable.

Creationists say time began with Creation, but cannot identify an exact date. This is inconsistent with the concept of an eternal God who existed <u>before</u> Creation.

When we look in the Books of Science, we find technology has advanced so far that Science can claim the ability to measure intervals of time with uncertainties less than parts per trillion. Details are in Exhibit 9.1, Type 1, § A.

One Solar Day corresponds to exactly one revolution of the Earth, which is the interval from when the Sun is exactly overhead to the next overhead. Unfortunately, the Earth has moved along in its orbit so it has to revolve a bit more on its axis to reach the next overhead. This longer interval is arbitrarily divided into exactly 24 hours of 3600 seconds each. The only means of realizing the SD is by a sundial. Dividing the total number of cesium oscillations in a SD by 86,400 yields the electronic definition of the second.

EST says actually sidereal time is used for observations. The sidereal day is the interval between successive crossings by a star past the meridian (the north-south line through the zenith at the observer's location). We count the number of cesium oscillations in a sidereal day and divide by 86,400 seconds to find that one solar second = 1.00273790935 sidereal seconds, and similarly for the lengths of the days. But:

- Why was this particular transition chosen? Are the cesium emissions really constant?

- How do we determine the accuracy? How accurate is the "known" frequency? We haven't been around for a million years. What standard is used? The Earth's orbit is an ellipse. From Kepler's Second Law there is about a 3.5% variation in the Earth's orbital velocity. This must affect the accuracy.

- Are the Relativistic effects of the Earth's gravitational and magnetic fields accounted for?

Note that we are not counting time, but the number of atomic oscillations. It is arbitrary, like the day.

Comment: Digging further, we find (**TDU** and **EST**, Physical Measurement) that atomic clocks are adjusted periodically to agree with sidereal time! So the atomic clock is merely a device for indicating <u>intervals</u> in an arbitrarily chosen phenomenon: the electronic transitions in cesium.

The year is defined in terms of the vernal equinox, the time when the Sun crosses the plane of the Earth's equator, making the number of hours of daylight equal to the number of hours of darkness. There is clearly a lot more to this than this brief description.

- At the equinoxes the length of the day is changing most rapidly compared to all other times.

- The light from the Sun takes about eight minutes to reach the Earth, so we see the Sun where it was eight minutes ago.

- The Sun has a large diameter, so is the crossing figured as the moment when the Sun is bisected by the plane of the equator?

The sidereal year is longer than the solar year by a factor of 1.00038773, or 1223.57 atomic clock seconds. Frequency is the number of occurrences of an event per second and is expressed in Hertz (Hz).

F. Mass and Weight

Mass is the quantity of matter in a body. "Weight" is the force exerted on that mass by gravity, and varies with the strength of its field. Whatever we "weigh" at sea level is less if we ride in an aircraft, and still less if we were an astronaut in space.

Example: At 9 Earth radii above the surface of the Earth, we would be at 10 radii from the center, and our "weight" would be 1/100 that at the surface.

Science cannot measure mass. It can only <u>compare</u> the quantity of matter in one object with that in another. The International Bureau of Weights and Measures near Paris, France preserves a prototype kilogram in the form of a platinum-iridium cylinder. This is an arbitrarily chosen volume of material that originally was supposed to be the quantity of mass in one cubic decimeter of pure water at 4° C.

Secondary standards (working standards) are calibrated against it, presumably by weighing on a beam balance. The comparison is actually between the gravitational force on the two standards, but since the acceleration of gravity is the same for both, being in the same location, effectively we have a comparison of masses.

The key to understanding the difference between mass and weight is to realize that a scale reading is only how many times greater a mass is than the standard.

G. Mole (Symbol: mol)

A gram mole is the amount of a substance that has as many elementary entities as there are atoms in 12 grams of carbon-12. The entities can be atoms, molecules, ions, electrons, or other particles. Instead of using the actual masses of atoms and molecules, an arbitrary scale of mass is set up using carbon-12 as mass 12.

	Actual mass, g	Relative mass
1 proton	1.67265 E−24	1.007276
1 C-12 atom	19.9268 E−24	12.000000

H. Force and Work

A force is a push or a pull. A sufficient force on a mass will overcome its inertia and accelerate it. The mass in this case is the inertial mass vs the gravitational mass in § F. Experimentally, they are numerically equal.

The work done in moving an object is measured by the product of the force exerted on the body and the distance it was moved. Work is given in mechanical units (newton-meters) or in thermal units (calories).

I. Law of Gravity

Science does not know what gravity is.

Aristotle believed all moving bodies needed to be pushed or pulled to keep them moving. Galileo (1564-1642) experimented with balls rolling on inclined planes and concluded that in the absence of opposing forces a body would move in a straight line forever. He even named this property inertia. He also found the velocity of the balls rolling down the inclined planes was subject to a uniform acceleration (**CP**).

1. Isaac Newton (1642-1727)

Newton's First Law of Motion is: "Every body perseveres in its state of rest, or of uniform motion in a straight line, unless it is compelled to change that state by forces impressed thereon." I really don't see any difference in this from Galileo.

His Second Law makes the critical advance by quantifying the relation between force and acceleration: "The acceleration, a, of an object is directly proportional to the net (external) force, F, acting on the object, is in the direction of the net force, and is inversely proportional to the mass, m, of the object." That is, a varies as F/m.

By the First Law, since the Moon travels in a <u>curved</u> path, the Earth must exert a force on it. From the period of the Moon and its distance from the Earth, Newton showed that the force decreased with the square of the distance. From Kepler's Laws the force exerted by the Sun on the Earth showed the same dependence on distance: $F_s = K_s M_e / R^2$. He made a remarkable and crucial inference: there had to be a similar force of the Earth on the Sun! It was given by $F_e = K_e M_s / R^2$. Then he made another remarkable inference: $F_s = F_e$. Thus, $K_s/M_s = K_e/M_e = G$, a universal constant, so $K_s = GM_s$, and $F_s = GM_s M_e / R^2$. In general, the force drawing two bodies together is:

$$F = G\, m_1\, m_2\, /d^2 = A/d^2,$$

where the m's are the masses of the bodies, and d is the separation of their centers of mass. It is true for point masses, and very nearly true for spherical bodies whose dimensions are small compared to their separation, as for the Solar System and the stars.

Amazingly, the constant G can be evaluated in a laboratory experiment. Further, if Body 2 is the Earth and Body 1 is any other body on the surface of the Earth:

$$F/m_1 = g = G\, m_2\, /d^2,$$

where g is the acceleration of the Earth's gravity, measurable in the laboratory as 980 cm/s^2. Since we know the radius of the Earth, we can solve for the Earth's mass, m_2. The constant g means that, neglecting air resistance, the velocity of every body falling to Earth increases 980 cm/s every second. Note that G is not called a property of space.

2. Interpretations

But there is something strange here. The Law of Gravitation describes the force <u>between</u> two bodies: each body is attracting the other. For a 1-gram mass:

$$
\begin{aligned}
g_1 &= Gm_1/d^2 \\
&= (6.672\text{E}{-}08 \text{ dyne-cm}^2/\text{g}^2)(1 \text{ g})/(6.378\text{E}08 \text{ cm})^2 \\
&= 1.640\text{E}{-}25 \text{ cm/s}^2.
\end{aligned}
$$

The force this exerts on the Earth is:

$$F = m_2 g_1 = (5.977\text{E}27 \text{ grams})(g_1) = 980 \text{ dynes}.$$

This is exactly as large as the force exerted by the Earth on the 1 g mass:

F   = (1 gram)(980) = 980 dynes.

Question: Science has yet to explain how a 1-gram mass can exert such a force on the whole Earth. If the Earth were alone in space, would it still exert its acceleration of gravity through space?

The Moon orbits the Earth and the planets orbit the Sun, not the other way around. It is a mystery why, even though the force is the same, it is the smaller mass that orbits the larger. I have not found any explanation of why this is so. Digging further, we find that Science acknowledges that the Moon does perturb the Earth's orbit.

3. Original value of g

A curious result is obtained if we express g in astronomical units. Let's use light-years (LY) instead of centimeters, and solar years instead of seconds.

g' = $(980.66)(31.557\text{E}06)^2 / (9.4605\text{E}17)$

    = 1.03227 solar LY/solar yr^2,

or about 1 LY/yr^2. In symbols,

g' = $g\, p^2/a$, where

g = 980.66 cm/s^2, p = 31.557E06 s/yr
and a   = cp = 9.4605E17 cm/LY.

Since the mass of the Earth has been increasing due to capture of meteorites, this suggests that its g value was originally 1. Seems mystical. This may be just a coincidence, because there does not seem to be any pattern in the data for the other planets in our Solar System. For the primordial Earth, with $g'_0 = 1$:

g_0 = $g'_0\, a/p^2$ = 950.00 cm/s^2.

The increase in g_0 to its present value cannot be due to a mass gain since the present rate of meteorite accretion is too small. So either a) the meteorite capture rate was much greater in the early years or b) the initial Earth had about the same mass as now.

But if the Earth had been closer to the Sun, p_0 and $(LY)_0$ would have been smaller. Equating g_{sun} and Earth's v^2/R yields $4\pi^2 R^3 = (Gm_s)(p_0)^2$, so R = 1.4646E13 cm and the average orbital velocity was 3.010E06 cm/s. Equivalent values are:

$$R = 0.160 \, [(LY/(100)^2], \text{ or about } 1/6 \, [(LY/(100)^2]$$

and $v = 1.004 \, c/(100)^2$.

> **Comment:** So we have some intriguing speculations. Science talks much about random events in our Universe. Why should there be such unusual relations of our spatial parameters to the velocity of light and to the light-year? Why should not the ratios be random, like c/71438, which would then have no significance? Note also that the ratios v/c and R/LY are independent of our measuring system.
>
> If someone were designing a model, he would use multiples of his standard of length. To a Cosmic Creator, the constant velocity of light offers a convenient yardstick for measurements. He could have made the Universe any size He wished. Apparently, He chose a scale factor of 1/10,000.
>
> Now He would not necessarily have used our standard of length of 1 meter. But if He used the above ratios on length and velocity, the values of R and v would be as observed. It is intriguing enough to follow further, in Chapter 12.

We next describe four interrelated quantities: energy, momentum, pressure, and temperature.

J. Energy

Science does not know what energy is.

I have not found a definition of energy that is as explicit as the definition of a circle, a solid, or a seed. Energy is defined in terms of its effects:

- It is (an invisible mysterious something that is) expended in doing work.

- Thermal (heat) energy causes things to feel "hotter" and so is interpreted as the mechanical energy of atoms and molecules.

Charles H. Peterson

We cannot identify a zero energy level in the Universe, so we can measure only differences in energies. Two of the forms of energy are of particular interest.

1. Kinetic energy (KE)

Kinetic energy is energy due to motion. From Newton's Second Law and the relation between uniform acceleration a, velocity v, and distance s:

$$F = ma = m[(v_2^2 - v_1^2)/2s]$$

$$Fs = mv_2^2/2 - mv_1^2/2 = KE_2 - KE_1$$

In words, if we apply a constant force F to a body while it moves a distance s in the direction of the force, its kinetic energy will have increased by the amount Fs, the work done on the body.

2. Potential energy (PE)

Potential energy is due to the position or arrangement of matter. From experience, an object in a gravitational field has potential energy. A rock above the surface of the Earth can do work if released because it acquires kinetic energy while falling to the Earth. E.g., we can drop a rock on a nut to crack it.

This is not strictly correct. It is the system consisting of the Earth and the object that has potential energy due to the separation between their centers of mass. Why is the loss of potential energy attributed to the smaller body?

We have no way of measuring the potential energy of a body at a particular point. We deal with changes in potential energy. The force on a body of mass m at a distance R from the center of the Earth is:

$$F = -(GM_em)/R^2 = -A/R^2$$

This is an attractive force in the negative direction, R being measured in the positive direction. If a body is in an orbit around the Earth, to move it a small distance ΔR radially away from the Earth from Point P_1 to Point P_2, the incremental work done on the body is:

$$\Delta Work = -(A/R^2)(\Delta R)$$

Because the force changes continuously with distance, we use calculus to integrate (add up) these increments of work for all the ΔR's between Points 1 and 2 to get:

$$W = A [(1/R_2) - (1/R_1)]$$

Since R_2 is greater than R_1, the work W is negative. It is convenient but arbitrary to choose as a reference level of PE a value of zero for the potential energy at a point P_2 infinitely far from Earth.

$$W = -A/R_1 = PE_1$$

Then its potential energy can only decrease (become more negative) as it moves closer to the Earth. To generalize, we drop the subscripts and say the potential energy at a point at a distance R from the center of the Earth is $-A/R$.

Keep in mind that R is the radial distance from the <u>center</u> of mass of the Earth. Even though we know that for a real body the mass and hence the gravity force is <u>not</u> concentrated in a point at the center, which seems like nonsense, that is the concept on which all the theories are based .

Comment: Because Science insists on such concepts, arguing that they permit predictions of the behavior of matter and energy, it is not surprising that a truly fundamental explanation of the basic elements of our Universe eludes our understanding.

Note also that the PE is always negative for an attractive force and goes to minus infinity at the center of the Earth.

Comment: Do we really believe that the potential energy gets this large? If actually true, to move any particle, no matter how small, from the center of the Earth to any other location would require an infinite amount of work.

The question may be academic because nobody is doing anything at the center of the Earth. Perhaps it helps understand gravity there. If one had to consider such situations, there would very likely be some other effect or force to be accounted for well before reaching the center. Keep this in mind when we discuss atomic structure and star formation.

K. Forms of kinetic energy

Kinetic energy ultimately is stored in atoms and molecules in several forms. A molecule has energy of translation (movement in a straight line) and, in some cases, energy of rotation. Internally there can be rotation of one or more groups about a chemical bond, vibration, and electronic motion. For a given temperature, the amounts in each form are not equal but vary about some average values for each kind of molecule.

Question: Why and how does the energy distribute in these proportions?

Science: (We have faith that) it always does. We do not know what controls the distribution.

The total internal energy is unknown and thus far indefinable, so energy can only be expressed relative to an arbitrary reference, or ground, state. The electronic energy presumably is not zero at 0 K. In the other four types, the energy is zero at absolute zero. The intrinsic energy of a mass expressed by the Einstein relation is a separate consideration.

L. Momentum

Momentum is the product of mass m and velocity v. Linear momentum is momentum in a straight line and has the units of gram-centimeters/second. More details are in Exhibit 9.1, Additive Relationships. Angular momentum is momentum in a circular path and is mv x r, the radius of the path, in units of g-cm^2/s. It is clearly different from linear momentum. It is also given by mr$^2\omega$, which shows the relation to angles. Omega (ω) is the angular velocity, measured in radians per second. One revolution in a circular path covers 360°, or 2π radians. So if n is the number of revolutions per second, ω is 2πn rads/s. For a given angular velocity, the angular momentum increases as the square of the radial distance from the center of rotation.

Going back to § J.2., the body falling from infinity to point R_1 has its momentum increased from 0 to mv. This violates Conservation of Momentum.

M. Pressure

Pressure is force per unit area. E.g., a 125-pound woman walking in shoes with 1/4" square heels will exert a pressure of one ton of force per square inch on the floor as her heel contacts the floor.

The pressure of a gas in a container is due to the enormous number of collisions of its molecules every second with the container walls. By the Law of Conservation of Momentum, the pressure is:

$$P = (2/3)[(N)(KE)_{av}] / V \text{ dynes/cm}^2,$$

where V = the total volume, cm^3
            N = total number of molecules present
and $(KE)_{av}$ = the average kinetic energy per molecule.

This includes several assumptions. The molecules behave like point masses; all collisions are perfectly elastic; and collisions with walls do not affect the two components of velocity parallel to the wall but do reverse the component perpendicular to the wall.

Pressure therefore is the result of the translatory kinetic energy of the molecules. It is a statistical concept, and while the formulas based on it may be mathematically correct, they are only approximate representations of reality (**KTG**). Multiplying through by V, the total volume, we get:

$$PV = (2/3)(N)(KE)_{av}$$

N. Temperature

The portion of the total energy that flows from or to a molecule because of a temperature difference is called "heat". Temperature is an arbitrary measure of the heat content of a substance. It, too, is a statistical concept, so if the number of molecules per unit volume gets too small, the concept breaks down.

Comment: Let's keep this in mind when there is talk of the "temperature" of outer space.

1. Temperature scales

Temperature may be indicated by the Centigrade scale, now called Celsius after its originator. Its scale is arbitrary: the freezing point of water is at 0°, and the boiling point of water is at 100°. The interval is divided, also arbitrarily, into 100 equal parts.

Experiments by Jacques Charles and Joseph Gay-Lussac in the latter part of the 18th century showed that the volume V_t of a gas when heated at atmospheric pressure was linear with temperature t:

$$V_t = a_1 t + a_2$$

Charles H. Peterson

Using the boiling point and the ice point (the temperature at which ice is in equilibrium with liquid water saturated with air) of water:

$$V_{100}/V_0 = (100\ a_1/V_0) + 1 = 1.366, \text{ experimentally.}$$
$$a_1 = 0.366\ V_0/100 = V_0/273.$$

Using the modern value:

$$V_t = V_0\ [1 + (t/273.15)]$$

At t = − 273.15°C, the volume is zero. In 1848 Lord Kelvin proposed that no heat would flow from a body at some very low temperature which he called absolute zero. The foregoing equation shows that absolute zero is − 273.15°C. The Kelvin scale starts at 0 K (kelvins). Science doesn't want to admit negative volumes are possible in our Universe.

> **Comment:** This assumes the linear relationship between temperature and volume at constant pressure holds all the way down to 0 K.

The Kelvin scale enables quantitative statements about the <u>availability</u> of energy at various temperature levels. Thus, 0 K does not mean the heat content of a substance at that temperature is zero, but that the heat available for transfer is zero. Since heat does not flow "uphill", the only way we could use heat from a 0 K source is to first transfer it to another medium at a temperature lower than 0 K, which as far as we know does not exist in our Universe.

> **Comment:** Or so I thought. This law of man held for only 100 years. **EST**, Negative temperature, is on temperatures below 0 K. They involve electronic and nuclear magnetic moments (spins). A positive K seems to mean that there are more spins in lower energy levels, while a negative K is the opposite. There are strange notions involved.
>
> • If there are as many spins in lower as in higher energy levels, the temperature is infinite.
>
> • Cooling from positive K values to negative values does not go through 0 K but through infinite K.
>
> • Negative K values mean bodies hotter than infinite temperature, not bodies colder than 0 K.

Well, it's going to take some more study to sort this out. For now, negative temperatures, we are told, apparently occur only rarely.

Question: How can infinite temperature be contained? The people who are working on nuclear fusion would like to know.

2. Perfect gas law

To continue with definitions of the quantities Science deals with, we need to know that experimentally many gases conform to the perfect gas law:

$$PV = nRT,$$

where   V   = volume, cm^3
          P   = pressure in $dynes/cm^2$.
          R   = the universal gas constant, 8.31441E07 ergs/(g-mol)(K)
          n   = the number of g-mols of gas
and   T   = the absolute temperature, kelvins.

The factor $n = N/N_A$, where N is the total number of molecules and N_A is the Avogadro Number, which will be discussed later in this chapter. For real gases, we insert another factor z, called a compressibility factor. It cannot be predicted; it may be correlated from experimental measurements as:

$$z \; = \; 1 + B(1/V) + C(1/V)^2 + \dots,$$

where B, C, etc. depend only on composition and temperature. Alternatively, using different constants,

$$PV = nRTz = n[RT + B'P + (C'/RT)P^2 + \dots]$$

For low pressures, terms in P^2 and higher are negligible.

3. Measurement of heat

Science arbitrarily defines the amount of heat needed to raise the temperature of 1 gram of water from 14.5°C to 15.5°C as 1 gram-calorie. The heat Q to raise the temperature of a mass m from T_1 to T_2 is:

$$Q \; = (c)(m)(T_2 - T_1) \text{ calories,}$$

where c is the specific heat, a proportionality factor that for the conditions just cited is 1.0 calorie/g-K. This constant can be measured under conditions of constant pressure to get c_p or constant volume to get c_v. For a perfect gas, $c_p - c_v = R$.

Comment: The specific heat is not constant, but varies with temperature. By putting the deviations into c, Science can retain the simple form of this "Law". Various empirical equations have been developed to describe the variation with temperature.

Comment: Since c_p generally increases with temperature, this means that the number of calories required to heat a substance from 120°C to 121°C is greater than that for 20°C to 21°C.

4. Relation to kinetic energy

Combining the perfect gas law with the PV relation obtained above in § M:

$$(KE)_{av} = (3/2)(PV/N) = (3/2)(1/N)[RT(N/N_A)]$$
$$= (3/2)kT = U \text{ (internal energy)}$$

where k is the Boltzmann Constant, $1.38066E{-}16$ ergs/(g-molecule K). This shows the direct relation between kinetic energy of translation of a molecule and temperature. For a gram-mol of gas:

$$U = (3/2)RT$$

When we heat a gas at constant volume, the internal energy changes by:

$$\Delta U = (3R/2)(\Delta T) = c_v \Delta T.$$

This interesting relation says that for a gas whose molecules behave like point masses, the specific heat measured under constant volume conditions is a constant, independent of temperature. Its value is 2.98 calories/(g-mol)-K. Experimentally, it holds for all the chemically inert gases (helium, neon, argon, krypton, xenon, and radon) to at least 3000 K (**CEH**, Table 3-174; **AIPH**, Table 4e-1).

Question: The range of atomic mass of these gases is 4 to 86. Apparently, any energy added all goes to increasing only their linear velocity. Yet we have images of spinning basketballs and planets rotating on their axes. We also read that the proton in the hydrogen atom not only has something called spin but it can be made to precess in a magnetic field. Why don't these atoms, which are not mathematical points, have rotational energy? Let's look into the mathematics of rotation.

5. Rotational energy

For a point mass m orbiting at constant velocity of n revolutions per second about a central point at a distance r from it, we have these relations:

- Angular velocity $\omega = 2\pi n$ rad/s

- Angular momentum J $= I\omega = mr^2\omega$ kg-m^2/s, where I is the moment of inertia of the mass.

- Kinetic energy KE $= (1/2)I\omega^2 = mr^2\omega^2/2 = J^2/(2I)$

Now consider a uniform sphere spinning in constant rotational velocity about one of its diameters. Each particle of mass in the sphere has the same mass, but the r's vary from 0 to the radius of the sphere R. By calculus, the result for the moment of inertia is:

$$I = 2mR^2/5 \text{ kg-m}^2, \text{ so}$$

$$KE = (1/2)(I)(\omega^2) = mR^2\omega^2/5 = J^2/(2I)$$

Niels Bohr postulated that in atoms the angular momentum of the electrons is quantized, i.e., it can take on only values that are integral multiples of $h/2\pi$. Let us assume that rotation of the roughly spherical nucleus would, if it occurred, also be quantized.

$$KE = [nh/(2\pi)]^2 [(1/(2I)]$$

Based on mean free path, the radius of the helium atom is 1.09E–08 cm. Its mass is 6.647E–24 g. Most of the mass is concentrated in the nucleus, which has a radius of about 1.1E–13 cm. Ignoring the electrons:

$$I = (0.4)(6.647E\text{–}24 \text{ g})(1.1E\text{–}13 \text{ cm})^2$$
$$= 3.217E\text{–}50 \text{ g-cm}^2$$

$$KE = [(1)(6.6262E\text{–}27)/(2\pi)]^2 [1/2I]$$
$$= 1.728E\text{–}05 \text{ ergs}$$

The translatory kinetic energy is:

$$KE = (3/2)kT = (3/2)(1.3807E\text{–}16)(288.15 \text{ K})$$
$$= 5.968E\text{–}14 \text{ ergs}$$

Comment: This suggests the helium atom does not spin because it would have to acquire such an enormous amount of energy: about 3 billion times that in all three translational modes. Heating to 3000 K would increase the translational energy by only 10x.

Question: How fast would the nucleus be spinning?

KE $= 1.728\text{E--}05$ ergs $= (1/2)I\omega^2$

$\omega \qquad = 3.278\text{E}22$ radians/s

v $\qquad = R\omega = 3.605\text{E}09$ cm/s $= 0.120$ c, c $=$ the velocity of light.

This is the tangential velocity at the surface of the nucleus.

If nuclear rotation is not quantized, then some smaller amount of energy would set it spinning. It does not seem impossible for the helium atom to rotate, but for now let's accept the heat capacity evidence.

The hydrogen molecule, on the other hand, is treated as a dumbbell shaped molecule consisting of two hydrogen atoms. Experimental data for hydrogen shows c_v increases with temperature but with plateaus at 3 cal/g-K for very low temperatures, at 5 for roughly ambient temperatures, and at 7 for temperatures above 5000 K with gradual transitions between these levels. The interpretation is that the value of 3 corresponds to translatory kinetic energy in three spatial directions, the 5 to the addition of two modes of rotational energy, and the 7 to two modes of vibrational energy. The transitions appear because the molecules have a range of individual velocities, hence can acquire the necessary energies over a range of temperatures.

6. Heat transfer

One mechanism for distributing the total energy among the different forms is the random collisions between molecules that eventually result in a steady state, or equilibrium, distribution at constant temperature. Since a hot gas radiates energy, there must also be photon transfers among the molecules.

Comment: Science does not know how these processes are regulated.

From the discussion of momentum in Exhibit 9.1, in a central collision of spheres of equal mass, Body 1 and Body 2 simply exchange velocities. The equations do not permit a slow molecule to speed up a fast one since they reflect experience (Conservation of Momentum: Equation 1 under Type 2,

§ C). In summary, at low pressures, the mean square velocity of the gas molecules determines the kinetic energy of the gas, which determines the temperature we measure. Together with the volume, it also sets the pressure.

O. Charge

Science does not know what charge is.

The encyclopedia says, "One does not (read: cannot) define it but accepts it as a basic property of elementary particles of matter." Matter can have either zero charge or a nonzero charge. There seem to be only two kinds of the latter, so they are arbitrarily called positive and negative. Science has never found in free space a smaller quantity of negative charge than that on an electron, or a smaller positive charge than that on a proton. And it has never found a way to destroy an isolated charge.

Around every charge is a spherical electric field that exerts a force measured in newtons/coulomb on a charge placed in it. What is an electric field? It is the mysterious presence electricity shows by its ability to produce some effect in the space near it. To visualize the field, Faraday postulated non-intersecting lines of force radiating from each charge. The direction of the force exerted by the field at any point is along the tangent to the line of force at that point.

Textbooks show lines leaving positive charges and entering negative charges as a conventional representation of lines of force. The radiating lines for the electric field from a positive charge terminate either on an opposite charge or go to infinity (**CVEP; UP**). The lines entering a negative charge must come either from a positive charge or from infinity.

> **Comment:** Let's think about this. Draw a circle on a piece of paper. Show a couple of radii. Now draw as many radii as you can between these two. Soon you run out of space on the circumference, i.e., the thickness of the line you want to draw is greater than the space available.

Now lines of force are merely a way of thinking. Still we have two possibilities. We can subdivide the circumference indefinitely, or we cannot. In the former, there is no end to the infinitesimal. In the latter, there is space between the lines of force. Think of the hairs on your head. However, this only exposes the inadequacy of the concept.

Comment: Science talks a great deal about symmetry in its theories, as if that puts them on sounder ground. Perhaps it does. But Science is inconsistent. It says our Universe is infinitely large yet insists that anything smaller than the Planck length (1.6E–33 cm) has no meaning. It would seem an elegant piece of symmetry to have our Universe include the infinitesimally small. This would be in accord with the view of some of the ancient Greeks, possibly including Aristotle.

What we do know experimentally is that at any radial distance from a charge, there is a force due to it that has the same magnitude regardless of which direction one chooses for the radius and which either attracts or repels radially another charge at that location. The distribution of force is continuous so there must be an infinite number of "lines" of force. This suggests that as long as we think of "lines" we are not thinking of what really is there.

Comment: It is also worth thinking about how something at infinity is tied up with lines of force from individual charges here. What is out there?

P. Magnetism

Science does not know what magnetism is.

Magnetism can either attract or repel. We have been aware of it for centuries as a property of lodestone, a natural form of magnetite. If a small needle-shaped piece of lodestone is put on a pivot, it will align itself with the north-south line through it. This shows the Earth itself is a magnet. But is the north-seeking end of the needle the north pole of the needle? Neither it nor the Earth came with labels.

The centuries-old agreement is that the Earth's North geographic pole is in the Arctic and its South geographic pole is in the Antarctic. The imaginary line connecting these two points is the Earth's axis on which the Earth rotates. The magnetic needle points to a location somewhat displaced from the North geographic pole that is called the North magnetic pole because it is near the North geographic pole. But this is apparently incorrect.

None of the references I checked told how the North magnetic pole was really identified as a south pole. There has to be some observation, perhaps experiments with moving charges. **UP** states that lines of force come out of the southern hemisphere, go through the air and enter the northern hemisphere. Also, lines of force come out of a <u>North</u> magnetic pole, which seems like an arbitrary decision.

We then say the north-seeking end of a magnetic needle must be a North magnetic pole since it is attracted to the South magnetic pole near the North geographic pole. **HEF** agrees and adds that the needle is said to point in the direction of a line drawn through it from its south pole to its north pole.

The source of the Earth's field is unknown. All magnetic fields are said to be due to moving charges. It is hard to imagine free separated charges within the Earth, but if such exist, convection currents in molten material and the rotation of the Earth might result in circulating currents and a magnetic field.

If we cut a bar magnet in two pieces, one cut end shows a north pole, the other a south pole. Science postulates that, as for electric charges, there should be isolated north and south poles (monopoles). None have ever been found. Modern theory attributes magnetism to electron spin, which will be discussed later.

Q. Waves

We have all seen some body of water lying quietly in a container: a tub, a reservoir, or a lake. A water wave is a disturbance in that body of water in which the surface moves up and down. If we throw a stone into a lake, we can see ripples move radially away from the point of entry in all directions.

Sound waves and water waves are mechanical waves that are propagated in (transported by) a medium: e.g., air or water. Neither medium moves in the direction of the propagation; only the disturbance is transmitted.

- Sound waves are compressional waves in a solid, a liquid, or a gaseous medium. The magnitude of the disturbance is propagated in the direction of travel of the wave.

- Surface water waves are transverse waves: the magnitude of the disturbance is propagated perpendicular to the direction of travel.

- A vibrating string also generates transverse waves. It differs in that its ends are fixed in position. It must be under tension, otherwise a wave could not be transmitted along it.

Question: This is interesting. Air, water, and metal bars transmit sound. Is there also some kind of tension in them? Or is the vibration of a violin string a different kind of phenomenon than transmission?

A simple wave can be pictured as a capital S turned on its side and then inverted. The picture would show a vertical displacement starting at zero, increasing to a maximum, then decreasing through zero to a minimum, and finally returning to zero. This is one cycle. If we drew a second cycle, that could represent the wave either at another point of observation or the cycle previous in time.

Waves have characteristic features. They have a mean (i.e., an average) value, which in the case of water waves is the level prior to the disturbance. Then there is an amplitude, which is the maximum displacement from the mean value. For the waves of interest here, the displacements in each direction are equal.

The distance between successive maxima is the wavelength. The frequency is the number of maxima per unit time, usually stated as cycles per second, or Hertz (Hz). This is named after Heinrich Hertz, a 19th Century German physicist, not the motor car rental agency. The frequency of audible sound is 20 to 16,000 Hertz, but sound waves have been generated up to 600 kHz. AM radio waves go as low as 530 kHz.

Many waves can be represented by the simple formula:

$$y = A \cos 2\pi vt = A \cos (2\pi t/T),$$

where A is the amplitude, t is time, v (nu) is the frequency, and T is the period. This shows that the displacement y varies from $+A$ to $-A$ and back to $+A$ over the period T because the cosine goes from $+1$ to 0 to -1 to 0 to $+1$. The formula also describes a simple harmonic motion (SHM) of an oscillating body.

The total energy associated with it is the sum of the kinetic energy and the potential energy, and is:

$$E = 2\pi^2 \, mv^2 \, A^2 \text{ in kg-m}^2/s^2, \text{ or joules,}$$

where m is the mass of the oscillating body. Note that this involves both the amplitude and the frequency.

R. Electricity and its fields

Science does not know what electricity is.

It is understood only by its effects. It is a mysterious something that has the power to cause effects that are believed to result from the presence, either static or moving, of particles carrying either positive or negative charges. Electricity can exert either an attractive or a repulsive force.

When an electric current flows in a straight conductor like a wire, there is both an electric field and a magnetic field around the wire. The fields pulsate in a regular manner and so waves appear to be travelling along the wire. The field is shown by the observation that two parallel wires will attract each other if each carries a current in the same direction, and will repel if the currents are oppositely directed.

Textbooks picture the electric field as vertical and the magnetic field as horizontal. This is misleading. The electric field is radial; the magnetic field is tangential to circles concentric with the wire. Both show a regular variation in amplitude (strength). They pulsate in phase, reaching their maxima together.

S. Electromagnetic waves

　1. Behavior

James Maxwell developed four equations to describe the behavior of electromagnetic waves. In 1864, he predicted that an oscillating circuit should generate such waves. Their velocity would be given by:

$$v^2 = 1/\mu_o\epsilon_o$$

where μ_o is the permeability of free space and ϵ_o is the permittivity of free space. The former is really only a proportionality constant for magnetic phenomena and the latter a similar constant for electrostatic phenomena. Both can be measured in the laboratory. The value calculated for velocity was:

$$v^2 = 1 / [4\pi E{-}07 \text{ henries/meter}]\ [8.854E{-}12 \text{ farads/m}]$$

$$v = 2.99795E08 \text{ m/s, or about the velocity of light.}$$

The units work out as follows:

1 henry/m	= 1 weber/ampere-m
	= [1 volt-second] /(coulomb/second)-m
	= 1 V-s^2/coulomb-m
1 farad/m	= 1 coulomb/V-m

$$1 \text{ henry-farad/m}^2 = 1 \text{ s}^2/\text{m}^2 = 1/(\text{velocity})^2$$

In 1887 Hertz was able to generate electromagnetic waves of a known frequency and to measure their wavelength to confirm that this was indeed their velocity. Thus, light was part of the electromagnetic spectrum.

> **Comment:** The constant μ_o now has an <u>assigned</u> value (**CVEP**). It is remarkable that v came out to be essentially equal to c. There is the nagging question as to whether this was fortuitous. The fact that two quantities have the same numerical value does not prove they are identical. No one has a theory as to why this relationship should exist. The value of ϵ_o is also assigned since it can be calculated from the velocity relation above and the assigned value of μ_o.

2. Wave energy

The field energy of an electromagnetic wave per unit volume of space at any given point has a form similar to that for mechanical waves (**CVEP**):

$$E_T = (1/2)\epsilon_o E^2 + (1/2)B^2/\mu_o,$$

where E = the strength of the electric field, newtons/coulomb
       B = the strength of the magnetic field, webers/meter^2
and E_T = the total energy, joules/m^3.

> **Comment:** To work out the units, a farad is a coulomb/volt and a weber is a volt-second.

Thus, the energies are proportional to the square of some quantity, which is at least analogous to amplitude in mechanical waves. However, we are also told that the energy of electromagnetic radiation (waves) is:

$$E = h\nu \text{ (joule-s)}(1/\text{s}) = h\nu \text{ joules}$$

Here the energy depends on the first power of frequency, not the square, and amplitude does not appear at all. See the equation at the end of Section Q.

> **Comment:** This seems inconsistent. Does Science consider amplitude in such waves to be constant and identical, hence included in h, or irrelevant?

This relation was postulated by Max Planck in 1903. He derived a law for radiation based on quantum mechanics (See Exhibit 9.1) that has been experimentally confirmed. Therefore, our questions about the postulate may be more about our lack of understanding of electromagnetic waves than about its validity.

3.   Electromagnetic spectrum

Many types of electromagnetic radiation are shown in tables of the electromagnetic spectrum.   Different sources give significantly different ranges for the frequencies. Table 9.1 presents figures from **HEF**, p 9-07, as a starting point. Some of the ranges overlap.

- Visible light

The range for visible light is the most definite in that it is set by the capability of the human eye. The maximum sensation in the eye is for yellow-green light. Easily visible light runs from 0.640 μ (red) to 0.480 μ (blue). The extreme range seems to be 0.7 to 0.4 μ. These wavelengths are found in Solar radiation. They are attributed to energy radiation by the outer orbital electrons in atoms moving between various orbits (energy levels).

- Infrared rays

These are longer wavelength light rays just beyond the visible red. They are attributed to oscillations and vibrations in atoms and appear as heat.

- Ultraviolet light

This is merely shorter wavelength light rays just beyond the visible blue. They are attributed to transitions of more tightly bound outer electrons.

- X-rays

Wilhelm Roentgen discovered that cathode ray tubes produced not only electron discharges from the cathode, but also a more penetrating radiation he called X-rays.   It has been attributed to electronic transitions by the innermost electrons of atoms.

- Gamma rays

Gamma rays are produced by nuclear reactions.

Cosmic rays are emanations received from our Sun and from outer space that can penetrate 16 feet of solid lead. They are described as streams of particles ranging from electrons to various atomic nuclei. They have mass and charge, so they are not electromagnetic radiation. The frequencies ascribed to them in lie in a wide range E22 to E34 Hz, and their kinetic energies go to at least 100 GeV. One GeV is the energy in an electron accelerated through a voltage difference of one billion volts.

4. Photons

Radiant energy is said to be quantized: the energy travels in little packets called photons. Science believes this since there is not enough matter in space to carry them like water carries waves. The photons are said to move in straight lines at the speed of light. Radiant energy is also absorbed or released in quanta, or not at all.

Table 9.1 Selected Types of Electromagnetic Radiation

Type	Wavelength, microns	Frequency, Hertz
Infrared	220[a]	1.36E12[b]
Visible light	0.700	4.28E14
Red	0.640	4.68E14
Yellow-green	0.556	5.39E14
Blue	0.480	6.24E14
Ultraviolet	0.400	7.50E14
Extreme UV	0.160	1.87E15
Unnamed	0.012	2.50E16
X-rays	0.16E-03	1.87E18
Gamma rays	12.5 E-06	2.40E19
Cosmic rays	0.56E-06	5.40E20

[ab] The wavelengths and frequencies range from the figure given to that for the next type of radiation: infrared goes from 220 to 0.700 microns.

Comment: If there is no transport medium in space, then photons somehow must 1) carry the energy of the radiation and 2) generate the oscillations. Photons after leaving the Sun would seem not to be connected to the Sun so the Sun cannot be influencing them. Each photon has been launched into space with a specific energy and is travelling at the velocity of light.

Question: What is oscillating? The electric current in use in the United States is generated by equipment that inherently produces a stream of electrons at a 60 Hz frequency. What and where is the mechanism for producing and regulating the oscillations in radiation in space?

Comment: Photons are also puzzling because they are considered to have zero charge since they are not deflected by electric or magnetic fields, and zero mass. Yet by Einstein's theories they are deflected by gravitational fields, which affect only mass. A gravitational field is due to mass, which is an arrangement of electrically charged particles.

Combining $E = mc^2$ with $E = h\nu$:

$$m = h\nu/c^2 = h/c\lambda,$$

since the velocity of propagation of radiation obeys another simple law, $c = \nu\lambda$, the product of a frequency ν (nu) and a wavelength λ (lambda). For a ray of red light of wavelength $\lambda = 6.4E{-}05$ cm:

$$m = (6.626E{-}27 \text{ erg-s}) / [(2.9979E10 \text{ cm/s})(6.4E{-}05 \text{ cm})]$$
$$= 3.45E{-}33 \text{ g, the imputed mass of a photon.}$$

This is less than 4 millionths of the mass of an electron, but it is not zero. And a nonzero value is consistent with the Einstein theory of gravity bending light rays slightly.

The wave nature of photons is difficult to visualize since there is no identifiable amplitude. The frequency of 100E12 Hz corresponds to a wavelength of 3.0 microns. A wave has no physical existence. Water waves are merely made evident by the movement in a physical medium, water.

Question: Does the 3 microns correspond to an average diameter of the photon? Is its diameter pulsating about this average value? This would require photons to exist in an enormous range of diameters. Radio waves would have photons 56 km in diameter. This seems ridiculous, not credible.

Question: How can we visualize photons of different energies? Remember they all travel at the same velocity, but their frequencies are different. Is the stuff of which they are made spread over the same volume of space?

Question: As particles, what holds each photon together? What regulates the process by which a photon metamorphoses into a wave and back again?

So one of the miracles of Science is the photon. It behaves like a particle but it has no mass, yet it has momentum and responds to a gravitational field. It has a frequency but no amplitude. It behaves like a wave, but does not require a medium.

5. Velocity of propagation

Science assumes that the velocity of light is constant and has a <u>defined</u> velocity of c meters/second. The meter is defined as the distance light travels in 1/c seconds. These definitions have us going around in a circle. We must fix the velocity of light or the length of a meter.

The best I can make of it is that in prior years the length of the meter had been fixed by the prototype meter in France. With this standard, many measurements were made of the velocity of light. Finally, the scientists agreed to reverse the procedure and set (define) the velocity of light as an average of four of them and then consider it an exact value. This value was 299.792458 million m/s. Presumably, a new meter bar was made to conform to this definition.

Comment: It would appear the reason for agreeing on such an odd number was that the meter would then be close to what it was earlier defined to be: one ten-millionth of the Earth's quadrant. For a radius of 6.3782E08 cm, the quadrant is 10.01885 million m. It is an important reason because the world's industry and commerce use the old meter standard and it would be very expensive to convert all the machines, nuts and bolts.

Comment: However, 300,000,000 m/s would have been such a nice round number. Pardon me for mentioning this, but it would have been $3 \times (100 \times 100)^2$. So perhaps the French were on track in 1793 by defining the meter in terms of a natural standard based on the size of the Earth. Despite the profound anticlericalism and out-and-out atheism in

France at that time, they were led to a unit of length based on a reference that many believe was created.

Question: Einstein's Theory of Special Relativity postulates c as an absolute value everywhere in the Universe and for as long as the Universe has existed. But there are qualifiers on the constancy: it is in vacuum, and in the absence of a gravitational fields. Since we now know space is not empty of matter, and since gravity is everywhere, how can we subtract out these effects to determine the "true" value of c?

It is convenient to discuss the meter here. The meter (m) is the distance light travels in a time interval t:

$$m = ct = [N_v U_v] [N_t U_t],$$

where N_v = the number of velocity units
U_v = the size of 1 velocity unit (e.g., 1 m/s)
N_t = the number of time units
and U_t = the size of 1 time unit (e.g., 1 s).

As of 1983, the meter is the distance traveled by light in $1/N_v$ seconds:

$$m = [(299792458)(1 \text{ m/s})][(1/299792458)(1 \text{ s})]$$
$$= 1 \text{ meter}$$

T. Electric current

Electricity flowing in the same direction through two parallel wires separated by a distance r produces a force tending to pull the wires closer. By experiment, the variables are related by:

$$F/L = [\mu_0/(4\pi)][2I_1 I_2/r],$$

where F = force, N
L = length of wires, m
I = current, amperes (A)
r = separation distance, m
and μ_0 = proportionality factor, henries/m.

This factor is called the permeability of a vacuum but it is only a proportionality factor. Its units are newtons/ampere^2. Its numerical value may be found by measuring all the other quantities in the relation.

It was decided to <u>assign</u> the value $4\pi E{-}07$ H/m to the factor and let the equation define the ampere:

$$I^2 = (4\pi/\mu_o)(F/L)(r/2)$$

This shows that the ampere is independent of the units of force, length, time and mass. It is nevertheless arbitrary because of the arbitrary force, $2E{-}07$ N/m. The ampere can be determined to about 5 ppm.

Philosophical Question

Let us pause to explore a philosophical notion arising when we ask how is Avogadro's Number, the number of atoms or molecules in a formula mass of a substance, determined. This was originally determined from Faraday's Constant, but now direct measurements on pure crystals are used.

Method: A crystal consists of a regular arrangement of its atoms into unit lattices. First, the density of the crystal is determined. Next, by X-rays the size and shape of the unit lattice can be determined, as well as the number of formula units in it of the chemical of which the crystal is made. All these quantities are related by:

$$\rho = [(Z/N_A)(M)] / V_u$$

where

ρ = Density, g/cm^3
Z = Number of molecules/lattice
M = g/g-mol

and

V_u = volume of unit lattice, cm^3/lattice.

Then:

$$N_A = ZM / \rho V_u$$

$$= (\text{Number/lattice})(\text{g/g-mol}) / [(g/cm^3)(cm^3/\text{lattice})]$$

$$= \text{Number of molecules per gram-mol}$$

This result seems to say the Avogadro Number depends on the unit of mass.

Explanation: If we used a reference mass say twice the size of the one in Paris and still called it a "kilogram", every mass we measure would be labelled as having half the mass it now does. Similarly for length. When we say something is 1 meter long, we mean the distance between its endpoints is exactly equal to that of our standard meter. If we doubled the standard for the meter and still called it a "meter", anything that was a meter would be a half "meter" long. What was a cubic centimeter would be 1/8th of a "cubic centimeter".

What it really says is that the <u>number</u> density of molecules is a constant, <u>independent</u> of our system of units. A gram-mol has N_A molecules. If we doubled the mass standard, the quantity of matter M in our new gram-mol would be doubled but so would the value of ρ, so the ratio M/ρ is unchanged. If we doubled the standard of length, the density would be divided by 8 but the volume of the lattice would be multiplied by 8. This means the number of atoms in a gram-mole is an absolute although we do not have an absolute standard of mass or length.

> **Question:** But what about the lattice size? Ah, you see, that is fixed. Fixed by Nature or fixed by God. There is nothing we humans can do to change its nature. And in the sodium chloride lattice there was, is and always will be one and only one molecule of NaCl. It is, how do we say, <u>absolute</u>.

> **Comment:** ABSOLUTE?! Listen to the screams of the rebellious! We have pulled the rug out from under them and their protestations that Science shows there are NO absolutes. Unlike the velocity of light, which is <u>postulated</u> to be constant, and is not if there is a gravitational field, the crystal lattice is there for all to see. Constant. Unchanging. Eternal. Thought-provoking?

> **Comment:** This is not to propose that we should worship crystals. Rather, Religion can say such objects illustrate the handiwork of God for us to be conscious of and to marvel at.

And what of the structure of the 103 elements? The structure of an electron or a proton? They, too, though unknown are absolute.

Exhibit 9.1
Examples of Relationships

Type 1: Constants

Second

In 1967, the 13th General Conference on Weights and Measures defined one second as the duration of 9,192,631,770 periods of oscillation of the radiation corresponding to the transition between the two hyperfine levels of the ground state of the cesium-133 atom.

The wavelength of this radiation is 3.26 cm, which is between the far (relative to the visible light range) infrared at 0.03 cm and the near FM at 108 MHz, or 278 cm. It is a marvel of technology that devices can be built to identify this wavelength precisely and to count the oscillations.

The principle is to match the frequency of an electronic oscillator with the resonant frequency of the atoms. Cesium-133 atoms are irradiated, causing electronic transitions as the atoms absorb and then release energy. A frequency synthesizer generates high frequency harmonics of a known frequency. These are subtracted from the unknown frequency (in this case, the cesium-133 oscillations) to create a beat frequency that is at a lower and countable frequency. This is in effect an atomic clock. It is said to be accurate to 1 second in a million years (32 seconds per quadrillion seconds).

Year

1 Solar year (SY)	= 1 tropical year.
1 SY	= 1 equinoctial year
	= 31.556 925 974 7E06 s in Solar year 1900 (**CVEP**, p 4; **CRC**)
	= 365 days 5 h 48 m 45.9747 s
1 sidereal year	= 31.55814954E06 s
	= 365 days 6 h 9 m 9.54 s

Velocity of light (Defined in 1983)

c = 299.79 245 8E+06 m/s, or 30 billion cm/s

Meter

A meter once was the distance between two scratches on a platinum-iridium bar. The 11th GCWM in 1960 defined it as the length equal to 1,650,763.73 wavelengths in vacuum of the radiation corresponding to an electronic transition in krypton-86. This has been supplanted; see § S.5. in the text.

Light-year (LY)

This is the distance that light travels in one year in a vacuum and in the absence of gravitational fields.

$$1 \text{ Solar LY} = (29.9792458\text{E}09 \text{ cm/s}) \times (31.5569259747\text{E}06 \text{ s/Solar Year})$$
$$= 946.052\ 840\ 5\text{E}15 \text{ cm}$$

$$1 \text{ Sidereal LY} = (29.9792458\text{E}09 \text{ cm/s}) \times (31.55814954\text{E}06 \text{ s/sidereal year})$$
$$= 946.089\ 522\ 0\text{E}15 \text{ cm}$$

Astronomical Unit (A.U.)

The mean distance of the Earth from the Sun by radar determinations by the Jet Propulsion Laboratory is 1.495 978 07E13 cm (92,955,807.27 miles). Since radar was used, it is probably the distance from the surface of the Earth to the point on the Sun where the density of the gas was great enough to reflect the radar beam. I don't know whether the radii of the Earth and the Sun were added to the radar measurement. The center of gravity interpretation will be used in this book.

Universal constant in Newton's Law of Gravitation

$$G = 6.6720\text{E}{-}08 \text{ dyne-(cm/g)}^2$$

There are only five significant digits in G, including the terminal zero. Therefore any calculation involving G is good to no more than five figures.

Standard value of Earth's acceleration of gravity

$$g = 980.665 \text{ cm/s}^2$$

The local value of g can be measured in very elaborate experiments to better than 1 part in 10 million.

Charles H. Peterson

Atomic Mass Unit (amu)

For a substance with a formula mass of exactly 1 gram atom, 1 amu = 1 g/N_A = 1.660567E–24 g/g atom.

Charge on an electron

C   = 1.602189E–19 coulombs

This is rather abstract. However, its reciprocal is 6.241461E18, the number of electrons in a coulomb.

It was first determined in an ingenious experiment in 1910 by R. A. Millikan. He measured the terminal velocities of micron-sized oil drops in air rising or falling between two parallel plates according to the balance of gravity and electrical forces on them. The drops had various static electricity charges. What Millikan found was the charges had a common factor, which he inferred was the charge on one electron. The rationale was that many of the oil drops picked up more than one electron.

Faraday's Constant

F   = 9.64845E04 coulombs/g-equivalent mass

F originally served to determine Avogadro's number by dividing it by the charge on the electron.

Avogadro's Number

N_A = 6.02204E23 atoms per gram mole

It is the same for every pure substance with a definite chemical formula.

Planck's Constant

h   = 6.62618E–27 erg-s.

Type 2: Additive relationships

Conservation of Mass

A general form of the conservation relation is:

Input – Output = Accumulation, or
I – O = A.

Illustration: In any chemical reaction, the total mass of the materials entering the reaction equals the total mass of materials produced, because no mass is accumulated. If we burn hydrogen gas (H_2) with pure oxygen (O_2) we get water:

$2 H_2 + O_2 = 2 H_2O$

If we burn 4 grams of hydrogen with 32 grams of oxygen we get 36 grams of water. The numbers reflect experience. However, we <u>can</u> say we are accumulating the component water, and depleting hydrogen and oxygen. Total mass is conserved because at any stage the total of water formed and hydrogen and oxygen left is a constant, 36 grams.

Conservation of Energy

This is the First Law of Thermodynamics. It has the same form as that for mass. Energy can be present in different forms, the common ones being thermal, kinetic, and potential. Being measurable in the same units (calories or ergs), they are additive.

Conservation of Momentum

Momentum is a vector quantity: it has both magnitude and direction. In general, momenta cannot be added or subtracted directly, but their components in three mutually perpendicular directions are additive.

Conservation of momentum means that the total momentum in a system at any time is equal to the total momentum in that system at any later time. This generalization is confirmed by experiment.

Suppose Body 1 of mass m_1 is moving in a straight line in some direction at velocity w_1 and Body 2 of mass m_2 is moving in exactly the opposite direction along the same line at velocity w_2. Their relative velocity is always the difference: $w_1 - w_2$.

Charles H. Peterson

By experiment, direct central impact of spheres always results in a reduction in <u>relative</u> velocity and the relative velocity after collision is in the opposite direction. Using v's for the velocity after collision:

1) $\quad v_1 - v_2 \quad = -e(w_1 - w_2),$

where e is an empirical factor for the elasticity of the impact. Perfect elasticity is for e = 1. For e = 0, we have a totally inelastic collision in which the colliding bodies move off as a single mass.

2) $\quad m_1v_1 + m_2v_2 = m_1w_1 + m_2w_2$

From these two equations we find (**HEF**, p 4-36, 37):

3) $v_1 = w_1 - (1 + e)(w_1 - w_2)[m_2/(m_1 + m_2)],$ and

4) $v_2 = w_2 + (1 + e)(w_1 - w_2)[m_1/(m_1 + m_2)]$

In general there is a loss of kinetic energy (KE):

5) $KE_1 - KE_2 = \frac{1}{2}[m_1w_1^2 + m_2w_2^2 - m_1v_1^2 - m_2v_2^2],$

which after some algebra becomes:

6) $KE_1 - KE_2 = \frac{1}{2}(1 - e^2)(v_1 - v_2)^2 [m_1m_2/(m_1 + m_2)]$

For e = 1, the loss of kinetic energy is zero, which reflects perfect elasticity.

Type 3: Multiplicative relationships

<u>Distance</u>

$$s = v\,t,$$

where s is the distance (meters) an object travels in time interval t (seconds) at velocity v (m/s). No proportionality constant is needed.

<u>Ohm's law</u>

This is a familiar law in the flow of electricity, and is an example of an equivalence relationship:

$$E = I\,R,$$

where E = voltage, volts
         I = current, amperes
and   R = resistance, ohms.

Here the proportionality constant is 1 because we can arbitrarily choose the size of the units of each of the variables to make the constant 1 volt/ampere-ohm.

Interconvertibility of mass and energy

This is the Einstein relationship and is an extension of the conservation of energy concept:

$$\text{Energy} = \text{Mass x Velocity of light, squared}$$

or, $E = m\,c^2$

This is a combination of Types 1 and 3. It is another equivalence relationship. It is not like saying 16 ounces is equivalent to one pound. It says rather that one thing, mass, can be stated in a completely different unit, energy, if we use a proportionality factor.

The theory says that this factor is exactly the square of the velocity of light. Although this is strange and hard to accept, it has been proven in a practical way by building nuclear reactors.

This only proves that the factor is <u>numerically</u> equal to c^2. Science still needs a theory as to why mass has this particular relation to energy.

Type 4: Power relationships

Kinetic energy

Here multiplicative and power relations are combined:

$$E = m\,v^2/2,$$

where E is the energy (ergs), m is the mass (g), and v is the velocity (cm/s).

Charles H. Peterson

Coulomb's Law

The force F between two electric charges Q_1 and Q_2 r meters apart is given by:

$$F = Q_1 Q_2 / [4 \pi \varepsilon_0 r^2]$$

F is in $[coulombs]^2/[farad-m]$
$$= \text{volt-coulombs/m} = \text{volt-ampere-s/m}$$
$$= \text{watt-s/m} = \text{joules/m} = \text{newtons}$$

The magnitude of the force between two equal charges is the same for two positives, two negatives, or a positive and a negative. The first two are repulsive; the third is attractive. The value of the constant is $\varepsilon_0 = 8.854187816E{-}12$ farads/meter, and is called the permittivity of free space. However, it is merely a necessary proportionality constant: force, charge and length have been defined independently of each other. It was also found convenient to arbitrarily separate out the factor of 4π. Its value is now calculated from the Maxwell relation: $c^2 = 1/\varepsilon_0 \mu_0$.

Stefan-Boltzmann Law

$$E = \sigma T^4$$

This describes the emission of thermal energy. Originally it may have been an empirical law, meaning it was derived from experience and observations. Boltzmann then derived it from thermodynamics. It follows from Planck's Law on integrating I, the intensity of the radiation (see Type 5.B), over all wavelengths from 0 to infinity (**PHT**, Ch.4).

$$E = \pi \int I \, d\lambda$$

$$= \pi \int [(2C_1/\lambda^5) / (e^x - 1)] \, d\lambda$$

$$= \pi \int [2C_1)(xT/C_2)^5 / (e^x - 1)](-C_2/x^2 T) \, dx$$

$$= \pi [2C_1)/C_2{}^4][T^4] \int f(x) \, dx,$$

where $f(x) = x^3/(e^x - 1)$ and the other quantities are defined under Planck's Law in Type 5 relationships. The integral yields an infinite series (**PHT**, p 74):

$$\int f(x)\, dx \quad = [\Gamma(4)][(1/1^4) + (1/2^4) + (1/3^4) + \ldots]$$

The sum of the series can be expressed in closed form by use of an incidental result from Fourier series analysis of the mean value of the square of the variable x^2 over the interval $-\pi$ to $+\pi$ (**AME**, p 201), and the value of the gamma function is 6:

$$\text{So } E \quad = 2\pi\,[C_1/C_2^4][6][\pi^4/90][T^4]$$

$$= [2\pi^5/15][k^4/h^3c^2]\, T^4$$

$$= \sigma\, T^4$$

$$\sigma \ = 5.670278E{-}05 \ (\text{ergs/s})/\text{cm}^2\text{-K}^4$$

$$= 5.670278E{-}08 \ \text{W/m}^2\text{-K}^4$$

It describes the emission accurately. Science has demonstrated its validity in the design of furnaces. The calculated value of σ can be confirmed in the laboratory, supporting Planck's Law (Type 5: § B).

Type 5: Exponential relationships

Radioactive decay

This is covered in Exhibit 12.1 in Chapter 12.

Planck's Law

This law was derived from quantum theory (**HHTF**, Ch. 14). The intensity, I, of monochromatic (i.e., single wavelength) radiation is measured in watts (W) per square meter per unit wavelength.

$$I \quad = (2C_1/\lambda^5) / (e^x - 1),$$

where h   = Planck's constant = 6.62618E–27 erg-s

$$C_1 \ = hc^2$$

$$= 5.955314E{-}06 \ \text{erg-cm}^2/\text{s} = 5.955314E{-}17 \ \text{W-m}^2$$

λ = the wavelength, m

$$x \quad = C_2/\lambda T$$

$$C_2 = hc/k = 1.438789 \text{ cm-K} = 0.01438789 \text{ m-K}$$

and $\quad$ T $= $ absolute temperature of the source, kelvins.

The wavelength at which the intensity is at its peak is found by differentiating I with respect to λ and setting the result equal to 0 (**PHT**, p 66):

$$(5 - x)e^x \quad = 5$$

Inserting the known value of C_2, by trial and error:

$$\lambda T = C_3 \quad = \quad 0.2897796 \text{ cm-K}$$

The maximum value is $C_4 T^5$; $C_4 = 4.095570E{-}06$ W/m^{3}K^5. We also have $C_1/\lambda^5 = hc^2/\lambda^5 = hvc/\lambda^4$ and $x = hv/kT$.

Planck found values for h and k by fitting experimental data to his intensity law.

Chapter 10. Technology: Concepts

Science's "explains" our Universe in terms of aggregations of matter and interactions of matter and energy. These rest on certain assumed concepts that are only required to fit the observations and be consistent with each other. To evaluate the Religion of Science, we must understand its concepts.

Interaction

Science has observed only two kinds of interactions in our Universe: direct contact and remote action.

- Direct contact means that two bodies or charged particles must collide in order to interact. Examples are two billiard balls, most chemical reactions, and nuclear fission by neutron bombardment.

- Remote action means a body or a charged particle affects another body or charged particle even if a vacuum separates them. Science still does not understand how this is possible.

Direct contact is by random collisions of particles. Two particle collisions are much more probable than three particle collisions. Collisions of four or more particles are extremely unlikely. Reactions that seem to involve many particles are usually due to a series of intermediate reactions that involve the collision of only two particles at a time, some of which may be charged fragments of molecules.

Direction of Action

A. Flow

Experience shows that water never flows spontaneously uphill. Science "explains" (read: describes) this by saying a pressure difference is required to make it flow uphill. E.g., we can drink a liquid through a straw because we reduce the pressure on the liquid in the straw below the pressure of the atmosphere on the surface of the liquid outside the straw.

Experience also shows that heat (thermal energy) flows spontaneously only from a "hot" body to a "colder" body. Science therefore speaks of temperature as a potential for producing heat flow, and describes the heat flow as a tendency to equalize potentials. Similarly, voltage (electric potential) difference makes charges move.

> **Comment:** Science cannot explain why this tendency should exist. Those who are philosophically inclined can meditate on its applicability. We have:
>
> > **Luke 12:48** "For unto whomsoever much is given, of him shall be much required . . . "

B. Chemical reactions

Chemical reactions <u>spontaneously</u> go in the direction of reducing the energy of the reactants. If we mix an acid with a base, the two chemicals will of their own accord react, releasing heat. But the reaction then cannot spontaneously reverse itself. It <u>can</u> be driven in the reverse direction, but only by adding energy, e.g., by electrolysis.

C. Nuclear reactions

Spontaneous nuclear reactions behave similarly. Whatever laws govern instabilities in nuclear structure, they operate toward stability by decay processes that eject mass, energy or both from the nuclei involved.

Equilibrium

This is a state of rest or balance among opposing forces, potentials or processes. E.g., our body temperature stays at 98.6°F because our body is losing heat as fast as it gains or generates heat. The Moon stays in orbit since the centripetal force from the pull of the Earth's gravity just balances the inertial tendency of the Moon to continue in motion in a straight line.

Entropy

A. Availability of energy

The statements in § B on direction of action are quantified in the concept of entropy. Consider that 1000 calories of heat in steam at 100°C is not the same as 1000 calories in ice at 0°C. We don't usually think of ice as containing calories, but everything contains calories relative to a lower reference temperature.

The quantity of heat is the same, but the availabilities are not. Steam in contact with ice will melt it, but ice in contact with water will not heat the water to a higher temperature. To melt ice, we must add 80 calories/g. If this is done at 0°C, the increase in entropy of the water is:

$$\Delta S = S_2 - S_1 = q/T = 80/273.15, \text{ or } 0.293 \text{ cal/g-K}.$$

If we extract this heat, the water will freeze, i.e. crystallize to ice, and the entropy of the water will decrease by this amount.

B. Definitions

None of the references I have consulted offer a simple definition of entropy. Because it enters in discussions of our Universe, it is useful to offer a condensed explanation. We need a few more definitions. The concept of entropy seems to have arisen from studies of how much useful energy could be extracted from a process. For these studies, we imagine a surface enclosing the process or engine (**CEH**). We have:

- Environment   What is outside the surface.
- System   What is inside the surface.

  o Closed   Energy but not matter can cross the surface.
  o Isolated   Neither energy nor matter can cross the surface.

C. Reversible process

Next we conceive of an ideal, or reversible, process. In it, an infinitesimal change in some condition will cause the process to proceed in the reverse direction since there are no losses to friction, etc. In a cyclic process, a working fluid undergoes various state changes, as from liquid to vapor, returning periodically to its initial state.

In a heat power cycle, a working fluid is used to obtain work from heat. It always has three elements: a high temperature energy source, conversion of some of the energy to work, and rejection of the remainder of the heat to a low temperature sink (**HEF**). For all reversible cycles, by Conservation of Energy:

$$Q_i = Q_o + W,$$

where the Q's are input and output heat quantities and W is the work done by the process. The efficiency E of the process is given by:

$$E = W/Q_i = 1 - Q_o/Q_i$$

In 1824 Sadi Carnot formulated three propositions regarding reversible cyclic processes that amount to the Second Law of Thermodynamics. He developed a proof (**TPCE**) that:

$$Q_o/Q_i = T_o/T_i, \text{ which is rearranged to}$$

$$Q_o/T_o = Q_i/T_i,$$

where T_i is the temperature of the input heat and T_o is that of the output heat.

Rudolf Clausius decided these ratios were an intrinsic property of matter, naming it entropy (S). He studied it in differential form:

$$dS = dQ/T$$

which means the change in entropy is the change in heat content of a system divided by the temperature at which the heat was transferred. He showed that for a cyclic reversible process, the entropy change is zero, which means the entropy remains constant. For an irreversible process, the change is always greater than zero.

D. Irreversible processes

In real processes, which are irreversible, thermal energy transferred to a lower temperature then has a greater entropy, i.e., is less available to do work. Generalizing, all matter and energy tends to move to the lowest level of availability. Rivers run downhill to the oceans, batteries run down, materials corrode, fertility of farm lands decreases with each crop, mountains get worn down . . . .

Comment: These observations suggest there is a Law of Nature that will tolerate differences but will not permit a gross disparity over the long term.

Some rush to the unwarranted conclusion that our entire Universe is running down. Since we do not know whether our Universe is open or closed, we cannot say whether our Universe is running down or not. If it is open, it might be acquiring energy from some outside source, or it might be supplying energy to some outside sink.

If closed and thermally isolated, it will reach a state of equilibrium in which there would be no net driving force to make processes go in a particular direction. It is equivalent to universal death, after all local disequilibria are equalized.

E. Entropy and probability

There is a Third Law of Thermodynamics, often stated as the entropy of a perfect crystal is zero at absolute zero Kelvin. This suggests entropy is somehow connected to the degree of order and probability.

> **Comment:** Perfect randomness might be a cloud of hydrogen gas, while perfect order might be a crystal of solid hydrogen. So the entropy of our Universe must be greater than zero and hence the probability of its existence is greater than zero. Perhaps I should modify Descartes' teaching from "I exist" to "I probably exist".

F. Absolute values

The literature says entropy, like temperature, pressure, volume and mass, can be given an absolute value. Energy, however, can only be measured relative to some arbitrary base.

> **Comment:** These seem to mean only that the first five quantities can take on zero values, whereas atoms at 0 K still have an unknown amount of energy. There is no reference point for zero length, but since a volume encloses at least one point, that point can be the reference point for zero volume. Our Universe is irrefutably 3-D, not 2-D or 1-D.

Heisenberg Uncertainty Principle

Since one of the purposes of this book is to offer specific information on how we know what we say we know, it is appropriate to go a little further than simply stating the Uncertainty Principle. The following is condensed from various mathematics and physics texts on the subject of waves.

In Chapter 9 (Waves), we noted that a wave could be expressed mathematically. The same equation applies also to a particle travelling in a circle at an angular velocity, ω (omega), which is in radians per second. Once around the circle is 2π radians, so we can write:

$$y = A \cos 2\pi vt = A \cos \omega t$$

131

Combining two sound waves of the same amplitude but slightly different frequencies will produce a tone of average frequency but whose intensity varies regularly between a maximum and a minimum. The maxima are called beats and occur at a frequency equal to the difference in frequencies of the two waves.

A pulse is an energy transient that starts at zero, goes to a maximum and then decreases to zero. Many waves can be represented by the sum of two or more other waves. Per **CVEP**, the range of angular frequencies creating a pulse T seconds long is:

$$\Delta\omega = 2\pi/T$$

CVEP then asserts there <u>is</u> a very important relationship between the energy of a particle and the frequency of its associated wave that earlier was called an assumption.

$$E = h\nu = h(\omega/2\pi).$$

The equation applies as well to pulses of energy:

$$\Delta E = (h/2\pi)(\Delta\omega) = (h/2\pi)(2\pi/T)$$

So T ΔE = h, which is Heisenberg's Uncertainty Principle in terms of energy and time. It says that if a pulse last for T seconds, the inherent uncertainty in its energy is ΔE. Similarly for changes in momentum:

$$R \, \Delta p = h,$$

where R is a linear distance measurement and p is momentum. **CVEP** then asserts that it is changing notation: R to Δx and T to Δt. I do not know if a rigorous proof of this exists or if these are merely "reasonable" extensions:

$$(\Delta t)(\Delta E) = h, \text{ and}$$

$$(\Delta x)(\Delta p) = h.$$

The interpretation is that there is an irreducible uncertainty in all measurements of position, momentum, energy and time. The more accurately one of these is measured, the more uncertain is its paired variable. **CVEP** says this principle is rarely of practical importance on any scale visible to the naked eye.

Fitzgerald-Lorentz Contraction

The Contraction concept opened the door to Esoterica. The ideas involved in it and those which apparently follow from it are not common sense ideas.

> **Illustration:** If we stand on a railroad track, we can plainly see that the rails converge and meet at some point in the distance. Pragmatism, or common sense, however, tells us that railroad cars run on these tracks and do not get narrower as they go down the track. Some other explanation is needed.

> Years ago I was told, I think in school, that it was because we have binocular vision. I did not question it because I had no need to. But why does the illusion persist when I take a picture of it with a one-eyed camera? I still don't know the explanation. But it is more important to be alerted to the fact that we cannot always trust the direct evidence of our senses.

A. Reference points

Once we believed what our eyes told us: the Earth was fixed in position and all other bodies in the sky rotated around it. It became part of our anthropocentric view of the Universe: viewing and interpreting everything in terms of our own self-centered experiences and values. It was an extension of the view each of us perceives from birth as an <u>absolute</u> truth:

> I am the center of the Universe and everything revolves around <u>my</u> perceptions, needs and wants.

After centuries of fearsome opposition from Religion, most now accept the fact that the Earth rotates around the Sun and the Sun itself moves through space. There is no longer any meaning to absolute position or motion because we have not been able to find any fixed reference point. If the distance between A and B is changing, A or B or both may be moving. Even the rate at which it is changing may be changing.

But there were more profound results. An early one was Galileo's conclusion that the laws of physics, such as the Law of Gravity, are independent of the uniform relative velocity of the observers (**CVEP**). Another involved the velocity of light.

B. The nature of light

There has been a long and still unsettled controversy over the nature of light. Its behavior can only be "explained" in terms of a dual nature: waves and particles, though never both at the same time.

- As a wave, light shows:
  o interference (Thomas Young, 1827);
  o refraction (Jean Foucault); and
  o diffraction patterns and polarizability.

- As a stream of particles, light:
  o casts sharp shadows;
  o shows a photoelectric effect; and
  o appears to have momentum.

Newton held to the particle theory. Foucault showed that the velocity of light in water was less than that in air.

C. Ether

During the 19th century, it was believed that there had to be a medium called an ether in space through which light travelled, if it was a wave.

Question: What was the nature of the ether? Was it stationary? Or did it move at the same velocity and in the same direction as objects near it? Was it isotropic (have properties that are the same in all directions)?

Newton had to accept the ether to explain refraction. There was also a question about the additivity of velocities. In the Galilean/Newtonian laws of motion, if someone on a train travelling at 60 mph walks in the direction of travel at 3 mph, relative to a stationary observer outside the train he is moving at 63 mph. Did this apply also to light?

Question: Suppose the passenger merely turns on a flashlight. It was known that light had a very high but definite velocity of propagation, c. If light was a wave, whether or not there is an ether, surely the total velocity of propagation as seen by a stationary observer is c + 60. Should we be able to affect the velocity of light in this way?

Suppose we put two flashlights base to base. The beams are surely separating at twice the velocity of light, since the velocity of light from each is constant. (See Addendum to Exhibit 10.1.)

D. Michelson-Morley experiment

Michelson and Morley (M/M) in 1887 tested the hypothesis that the Earth moved through a stationary ether. Some consider the M/M result a crucial finding leading to the birth of modern physics. To grasp the subtleties of the Lorentz Contraction we need to have in mind the nature of the M/M apparatus (**CVEP**).

> **Description:** Consider a square and its diagonals. Label the corners clockwise in the sequence A, B, C, and D and call the intersection of the diagonals O. A half-silvered mirror was mounted at O to split a light beam from A to send one half to B and the other half straight ahead to C. A mirror at C returned this half to O. A mirror at B returned the other half to O to join the half from C. Both then went on to D to record the interference pattern that should have resulted if there was a difference in velocities. The path lengths were identical. The portion of the path along AOC was parallel to the Earth's velocity while the portion along BOD was perpendicular to it.

> **Comment:** The crucial part of the experiment was to compare the travel time for the path OBO with that of OCO. Note that the light beam in each of these paths travelled first in one direction and then in exactly the opposite direction.

The principle used to make visible the effect of travel direction was interference. If two light beams really were waves and were exactly in phase, combining them would make a brighter beam, or, in this case, a beam as bright as the original. If they were exactly out of phase, the beams would cancel to produce darkness. It was believed that beams of particles do not show such effects. There is an incidental question.

> **Question:** We have been told that we cannot create or destroy energy, and that every wave carries energy in accordance with Planck's Law $E = h\nu$. A photon in a ray in the red portion of the spectrum that has a wavelength of say 6.4E–05 cm has an energy of (6.6E–27 erg-s)(c/λ), or 3.1E–12 ergs. In cancellation, where does the energy of the two canceling photons go? If two cars collide, their kinetic energy is

reduced to zero being converted to heat (thermal energy), noise (sound energy), and damage (mechanical work). Has Science suspended the Law of Conservation of Energy?

The setup was a bit more elaborate. The length of the OCO path could be adjusted by small amounts by a calibrated micrometer screw. The mirror at B could be tilted at a slight angle. Together, the experiment could begin with an image showing interference bands.

An equation was developed from G/N laws to predict the band shift when the apparatus was rotated 90°. The derivation is not entirely clear; more research is needed to resolve questions about it. It used two frames of reference. One was fixed in the ether, and thus moved at the same velocity as the ether, if any. The Earth and the M/M apparatus were in the other and moving at a fixed velocity relative to the fixed frame.

The light source on the apparatus was stationary with respect to the Earth. The velocity of its light beam when directed parallel to the Earth's path should have been larger by the Earth's orbital velocity of 30 km/s. The difference of 100 parts per million was enough to have been detected. A shift as small as 0.001 x the wavelength of light, or about 0.0000006 mm, could have been detected. How? There was passing mention of the use of multiple reflections.

Whether we agree with the equation or not, the fact is that no shift at all was detected. The question was: how should this be interpreted?

- G/N laws of motion were insufficient to describe the propagation of light; or,

- There was no ether, within the sensitivity of the experiment.

   Comment: The logic was that, since there was no observed effect (no change in the velocity of light), there was no cause (no ether).

Now Science believes the laws it has discovered are universal. It also has absolute faith in logic. The above logic has a positive corollary: for every effect, there must be a cause. Let us therefore apply this corollary to another observation.

Example: We observe a complex Universe (effect); therefore it must have a cause. We find, however, that this inference is resisted fiercely and reflexively, like a knee-jerk, in violation of Science's own logic.

This is indeed strange. It means that the Laws of Science are to be applied selectively. Who is the Lord Chief High Selector? It further means that some of the teachings of Science are to be regarded as holy, unquestionable, and worthy of faith so strong as to be fanatical.

The most acceptable conclusion was that the G/N concept of velocity addition did not apply to light. It is not certain that the experiment also disproved the existence of an ether. Lorentz retained the notion in his explanation of the non-additivity of velocities (**STR**). Einstein later discarded the ether as unnecessary although he could not explain light propagation in a vacuum. The experiment had at least revealed a greater mystery: if G/N laws did not apply, what was the cause of the negative result?

- Something changed the velocity of light.
- Something changed the lengths of the light paths.
- Some unknown phenomenon occurred.

E. Lorentz contraction

G. F. Fitzgerald and H. A. Lorentz independently in 1895 proposed an explanation of why the M/M experiment was negative. Through what may appear to be a mathematical sleight-of-hand, Lorentz later derived equations for a light beam moving at constant velocity v relative to a stationary observer to show why its velocity remained at c. **STR** says they were derived theoretically from the electromagnetic properties of matter. Perhaps there is somewhere in the literature such a definitive derivation.

There was an unwelcome condition: the explanation required acceptance of some radical concepts. To a stationary observer:

- A meter stick traveling at velocity v in the direction of its length would <u>appear</u> to be shortened by a factor F given by:

$$F = [1 - (v/c)^2]^{0.5}.$$

- A second on the moving body would <u>appear</u> to be lengthened by the factor 1/F.

- The factor F was the same for all bodies and did not depend on the material of which those bodies were made, just as for the Law of Gravity.

The path lengths for the light beams were the effective meter sticks. The path of the light beam travelling perpendicular to the Earth's motion was assumed to be unaffected, while the path parallel to the Earth's motion had a contraction, so the travel time for this path was shortened. This was exactly offset by the lengthening of the second in this path so that fewer seconds were ticked off. The same result would be obtained for velocity in the reverse direction.

Science accepted these bizarre notions rather than say something unexpected happened to light beams in moving through space.

Comment: The lay reader has a difficult decision to make. Should he believe that moving objects experience an <u>actual</u> contraction in length and a lengthening of their time intervals, or are these only <u>apparent</u> effects to a stationary observer? The question arises because some writers say "appears to the stationary observer" and others say "the contraction is".

Question: Consider the "stationary" observer.

- The mean radius of the Earth is 6378 km, so all of us on the surface of the Earth are being whipped around at about 0.5 km/s (1120 mph).

- The mean distance from the Sun is 150 million km, so the Earth is travelling around the Sun at an average velocity of 30 km/s (67,000 mph).

- Our Sun has an orbital velocity around our Milky Way galaxy. Its distance from the center of our galaxy is estimated as 30,000 LY and its period as 250 million years. Its mean velocity is about 225 km/s (503,000 mph). **TNU** says 1 million mph.

- Our galaxy may have some velocity through our Local Group of galaxies.

- Our Local Group of galaxies is moving toward a Virgo cluster of galaxies at 780 km/s (0.26% of c).

For velocity much smaller than c, the factor F is very close to 1.0. Here are some F values, expressed as a difference from 1.0; the ratios are relative to Earth.

	v, km/s	1 − F	Ratio
Earth	30	5.01E-09	1
Solar System	224	2.82E-07	56
Local Group	780	3.38E-06	676

Since the directions of these velocities are not necessarily perpendicular to that of the Earth, any contractions due to the Earth's velocity might be overwhelmed by those due to these larger velocities. Of course, this does not prove a contraction does not exist, but it may be an alternative explanation of why no effect was found. Also, the mathematical form of the effect might be different from that proposed by Lorentz.

Whatever is doing the adjusting to velocity has to adjust to several different velocities. Any determination of c by our "stationary" observer has built in effects of all these velocities in different directions. Also, the velocities are on curved paths, which, we are told, means accelerations are involved. By the General Theory of Relativity, accelerations affect the velocity of light. But maybe the effects are tiny . . . .

Another factor is that we think of the Earth's orbit as a circle around our Sun. Well, of course, it is slightly elliptical. But that is not the whole story. It is really an open curve made up a series of shallow saucer-shaped curves.

F. Derivation of the Lorentz Contraction

The Lorentz equations have been applied to a variety of phenomena, so we must understand their basis thoroughly. We must follow the steps of the derivation to see if each formally follows by accepted mathematical operations from the previous one. We must understand what the equations represent and decide whether any reasoning involved is valid.

The Book of Einstein (**RSGT**), which is a popularized version of what appears to be one of the principal books of the Bible of Science, gives one version. I would have preferred to just summarize this but after much study some questions remain.

- The derivation is simplified by using the reference point x = 0 instead of an arbitrary point x = x_1. **RSGT** asserts that the same results will be obtained. The derivation is expanded in Exhibit 10.1 to show that this is at least not obvious.

- A term appears in one equation that calls for a velocity greater than that of light. No mention, interpretation or explanation of this is given.

- The derivation introduces parameters that are called constants, yet as shown in Exhibit 10.1 they are functions of the relative velocity.

- Most importantly, there appears to be a violation of the rules of algebra.

We are left with no credible basis for the Contraction. We can say it is empirically true, meaning that, while we can't derive it from basic considerations or a mechanism, it seems consistent with certain observations.

G. Sizes of the corrections

Table 10.1 shows Relativity corrections for a larger range of velocities. For even a 1% correction, velocities must be greater than 14% of the velocity of light, or 42,090 km/s (26,150 mph). The farthest galaxies are supposed to be moving at 60,000 km/s (20% of c). Thus, Relativity effects are significant only for phenomena occurring at very high velocities: electromagnetic radiation and high energy physics.

H. Contraction effects: one dimensional

Besides the linear contraction effects, **RSGT**, p 51, also states that Lorentz hypothesized that the <u>form</u> of an electron contracts in the direction of motion and that this has been confirmed with great precision.

Comment: This is very interesting because I have not found any clear-cut agreement as to the diameter of an electron. The range of estimates exceeds a factor of 5 million (See Ch. 11).

Question: What happened to the wave model of an electron? **RSGT**, p 50, says that the negative electrical masses making up the electron were held together by a force able to overcome the repulsion of like charged

Table 10.1 Relativity Expansion Factors, 1/F

v/c	1/F
0.1404	1.01
0.4166	1.10
0.87	2.03
0.98	5.02
0.995	10.01
0.9999995	1000.

particles. Einstein is clearly holding to the particle nature of electrons as well as implying the electron is made of smaller particles.

Comment: What does "the form contracts" mean? If we squeeze a rubber ball, its form contracts but no mass is lost. The ball bulges out in other directions. Is the same true for the deformed electron?

Question: Recall the Doppler effect: the pitch of a whistle from a moving train drops as it passes by a stationary observer. To someone on the train, the pitch is unchanged. It only appears to change.

I. Contraction Effects: Three dimensional

The motion of a body moving at an arbitrary angle to ours can be resolved into three components. One will be parallel to ours, and the others will be perpendicular to ours. There should be an apparent generally different contraction in all three. Relativity says we will observe only the parallel contraction. But the object we are observing will be distorted in all three dimensions to other observers.

From the viewpoint of that body, <u>we</u> are continually being distorted various amounts in three dimensions. But <u>we</u> don't perceive ourselves as distorted. We are told that the velocities available to us are small and the effects are also small. This is not satisfactory since we want to understand the notion.

Let's follow this further. From the table above, if one travels at **87%** of the velocity of light, all his linear dimensions parallel to the direction of travel will be one-half of those when at rest. What does this mean in practice?

- It is only apparent to a stationary observer.

- Or, it is real: all of the traveller's tissues are squashed. Ordinarily, this means a sphere becomes prolate, something like a football. Matter is not lost, only redistributed. His thickness in the direction of travel will be halved, he would become pointy-headed, and perhaps 40% taller and 40% wider. Probably dead.

Question: We know enormous pressures must be exerted to compress solids or even liquid water. Is space capable of exerting such pressures?

- How can it affect time intervals? It is not enough to assert that a dimension is shortened in the direction of its velocity. What <u>causes</u> the contraction?

Comment: Nowhere in the derivation of the Contraction is there any discussion of what happens to volume, and hence the quantity of matter, when a body contracts. Perhaps the derivation was intended to apply only to massless photons or other electromagnetic radiation. These questions also suggest that the effects are only apparent.

His mass will have doubled. Not his "weight" but the <u>quantity</u> of matter in him. This is even more puzzling. If the increase is real matter, what is its source?

Comment: It seems more reasonable to consider that his mass is not being increased but his inertia is. Remember, Science does not know why matter has inertia.

Or, half of his molecules have been removed by some mysterious mechanism. We have to consider this possibility because as his velocity increased to c, he would become denser than a neutron star or a black hole. They would be restored by some mysterious mechanism on decrease in velocity. So he would die and be reincarnated. Science will probably balk at this because it sounds too religious.

J. Contraction effects: measurement of time

RSGT has an illustration showing that the interval of time between 0 seconds and 1 second in the moving system <u>appears</u> to the stationary observer as increased to 1/F stationary seconds.

Comment: Let's go back to basics. We have:

1 day = (24)(1 hour) = N x U,

where N is the number of time units and U is the size of 1 unit. Is the unit halved (v = 87% of c), or will one's heart beat drop to 36/minute? Since we don't know what time is, this notion is harder to critique.

Question: The earlier question was: "Should we be able to influence the velocity of light by propagating it from a moving source?" Shouldn't we ask, "Should we be able to influence the electronic processes in an atom merely by giving that atom a velocity?" Didn't Galileo conclude that the Laws of Nature are independent of velocity? Bringing the same skepticism to this as Science brings to Religion, what tells an atomic clock to make longer ticks? If there is a real change in tick length, what changes in the atoms of the clock and what automatically regulates the amount of the change?

Comment: It is argued that observers on a spaceship travelling faster than we on Earth would experience time passing more slowly. However, relative to them <u>we</u> are travelling at the greater velocity. So we are the ones who would live longer. If this is not true, then Relativity is <u>not</u> symmetrical. Symmetry seems to be built into the equations. And Science seems to regard Symmetry as another Law of Nature, if not a major god.

In fact, we don't need observers. Let us fire off a tiny projectile into space with an enormous velocity. Give it a system for constantly ejecting a small mass to give it an acceleration to further increase its velocity. Relative to it, we on Earth are travelling at an enormous velocity. Therefore we would all live longer. Some object this is not valid: the moving person stays younger. But who is moving? Motion is relative.

Comment: However, since the Earth is moving in constantly changing directions, the traveller would have to be constantly adjusting his velocity by accelerations and decelerations to hold constant velocity relative to Earth. The equations are not valid for examples involving accelerations.

K. Contraction effects: value of v

To use F, we need a value for the velocity v. How is it determined? Since v was introduced improperly in the derivation, we have no theoretical basis for v. The measurements by stationary observer A yield an apparent v. We need the true value of v to correct the apparent length of say a moving meter stick as well as the apparent time interval involved.

Relativity

A. Special Theory

Einstein built on the work of Lorentz and others to propose a Special Theory of Relativity with two basic postulates (assertions):

1) The velocity of light in a vacuum and in the absence of gravitational fields is constant; and

2) Physical laws of nature have the same mathematical form when referred to any inertial system of coordinates.

What 1) seems to mean is that observers in two separate systems moving at constant velocity relative to each other will find the same value for the velocity of light when measuring in their own system, or when observing it in the other's system. Postulate 1) is certainly inapplicable at the outset because everywhere we go we find gravitational fields, and even space is no longer the vacuum once supposed.

Postulate 2) means that if the form of a law under study changes when referred to a different coordinate system, it is not a Law of Nature. It is probably a necessary but may not be a sufficient condition. But what is the proof of 2)?

B. General Theory

This asserts that the effects of a gravitational field are indistinguishable from an acceleration. For an observer subject to a uniform acceleration, light travelling perpendicular to the direction of the acceleration would appear to travel in a path curved in a direction opposite to that of the acceleration.

Comment: The claim is simplistic. A mass has a gravitational field which produces an acceleration. Accelerations can be produced by other means which do not also show significant gravity effects. E.g., the launchers for satellites produce greater accelerations than that of Earth's gravity, but they have gravitational attractions far less than that of the Earth. Further, since we do not really know what gravity is, it may have other properties we have not yet uncovered.

A gravitational acceleration is also distinguished by the fact that it reaches to infinity whereas the acceleration produced by a rocket launcher does not.

Observers in each of two systems moving at constant velocity relative will also find that Newton's Second Law does not remain invariant (**CVEP**) when an acceleration is involved. The Relativistic corrections are complicated. **CVEP** dismisses the problem by asserting that acceleration does not play the same role in calculating force as in the case of low velocities and generally the velocity is small and so the correction is negligible.

> **Comment:** What is that role? These arguments are unsatisfactory. E.g., observations of curved high velocity particle tracks under the influence of magnetic fields must surely involve accelerations. One of these observations is supposed to be supportive of the increase in electron mass predicted by the Lorentz Contraction.

However, as a start in making the drastic change needed to fit Newton's Second law into the Theory of Special Relativity, Einstein chose to assume that the Law of Conservation of Momentum was valid (**CVEP**, 293).

C. Experimental confirmations

Although we can ask questions like the above, several experimental observations are said to agree with predictions from the theory. Consider six of them.

1. Bending of starlight

It was predicted that star light passing near the Sun would be bent by the Sun's gravity toward the Sun, so the star would appear to be displaced outward away from the Sun. Comparison of the positions of several stars near the Sun during a solar eclipse with their positions several months before or after are said to confirm this effect.

Comment: The differences in positions amounted to a few hundredths of a millimeter on a photographic plate (**RSGT**, p 128). What does that look like? The period at the end of this sentence is about 1/2 of a millimeter in diameter. We are talking about, not half, nor a tenth, but a few <u>hundredths</u> of that. It is indeed remarkable that Science stands in awe before such minute differences and worships them as real, especially since a number of unspecified corrections had to be made to the data. The actual angular displacements were in the range 0.08 to 1.00 seconds of arc. Presumably, these were <u>calculated</u> from the black marks on the photographic plates.

Question: How did the variations in the Sun's gravity affect the light beams as they passed from deep space past the Sun? The Sun's gravitational field varies strongly with distance. What about dark matter and the curvature of space?

Comment: It is said the effect has also been reported by radio astronomers.

2. Slowing of clocks

Another bit of evidence offered in support of Relativity is the reported confirmation of the slowing down of time in clocks flown around the world in a jet plane. I do not have a reference, but suppose the plane flew at 600 mph in a path 24,000 miles long. This would take 40 hours, or 144,000 seconds. The Relativity correction factor on time would be $F = 1 + (4.00243E{-}13)$, so the reference clock on Earth should show an elapsed time longer by 58 billionths of a second. Apparently technology can detect this difference with an uncertainty of only about 10%. From Exhibit 9.1, the clock accuracy is 32 seconds per quadrillion seconds, or 5 billionths of a second in this case.

Comment: The path of the moving clock was a circle, not a straight line. The Contraction was derived for velocity in a straight line. Motion in a circle involves an acceleration that must be accounted for.

Question: How was the plane refueled? If the plane landed, how was this accounted for in the elapsed time?

Comment: In contrast, we have:

Psalms 19: 1-3 "The heavens declare the glory of God; and the firmament sheweth his handiwork. Day unto day uttereth speech, and night unto night sheweth knowledge. There is no speech nor language, where their voice is not heard."

Translation: We have only to look about us, without telescopes and cameras, to be reminded of the colossal forces operating in our Universe over which we have <u>absolutely</u> no control.

- We can build skyscrapers, but we can not stop an earthquake or a volcano.

- We can go to the Moon, but we cannot start or stop a hurricane.

- We can murder each other, but we cannot create a single living cell.

Clearly, there is a gross distortion of perspective when we crow over 60 billionths of a second as a grand achievement of Man, and ignore the grandeur in our Universe.

Comment: Consider next that 600 mph is only 1/6 of a mile per second vs the contributions of the five other velocities we are subjected to. Have the effects of all these been taken into account?

Since they lie in different directions, there is a different lengthening of time in each direction. How does one figure the combined effect, if that has any meaning? For the largest, F is 0.9999966, which differs from 1.0 by 3 parts per million and would increase the elapsed time by about 0.5 s, or 8 million times that due to the plane velocity.

Comment: Even if the critique of the <u>derivation</u> of the Lorentz Contraction offered here is valid, the <u>formulas</u> might still be applicable, fortuitously.

3. Gravitational redshift

Energy that forces its way into an atom excites the atom. The added energy requires one or more of its electrons to move farther from the nucleus. Yet something in the makeup of an atom forces it to reject the extra energy soon and the electrons move back closer to the nucleus.

Question: What is this something and from what does it arise?

147

The energy released shows up as spectral lines on spectrographic plates. **RSGT**, Appendix III, asserts that if the atom is in a gravitational field, say on the surface of a star, the photon frequencies will be redshifted by an amount dependent on the strength of the field. I intended to merely summarize the derivation given, but again I found many questions. The development of the relationship is presented in only two pages and the logic is difficult to follow because:

- It jumps between the effects of linear motion, rotating discs, and gravitational fields.

- It does not provide the basis for some steps.

- It leaves out intermediate steps.

It also does not make clear whether a change in the length of a time interval is real or only as seen by a stationary observer. An effort to reorganize it is in Exhibit 10.2.

Photons leave a star at constant velocity as electromagnetic radiation, so the only parameter that can change is frequency. The redshift is said to exist for white dwarfs, but is undetectible for our Sun: about 2 parts per million.

$$GM/c^2 \quad = (6.672E{-}08)(1.989E33 \text{ g})/(c^2) = 1.476E05$$

$$GM/Rc^2 = 1.4766E05 / (6.960E10 \text{ cm}) \quad = 2.21E{-}06$$

Comment: It is not clear why the amount of the redshift is calculated for a point on the surface of the Sun. The Sun's gravity does not end there. If the photons arriving at Earth have lost energy, the loss should be calculated at the Earth's surface. Perhaps an answer is that the photon is released during an interaction between the Sun's gravity and an atom on the surface of the Sun. Once the photon has left the atom this consideration should no longer apply.

Question: How is the total red shift from stars split into the velocity effect and the gravitational effect? There are also gas clouds in space that redshift starlight.

We shall now use this result to find the effect of Earth's gravitational field on frequencies, specifically, to the effect on the slowing of clocks. Let's assume the plane discussed in § C.2 flew at a constant 30,000-foot altitude. From above:

$$1 - (v_m/v_s)_1 \quad\quad = \quad A/R_1$$

$$1 - (v_m/v_s)_2 \quad\quad = \quad A/R_2$$

$$(v_m/v_s)_2 - (v_m/v_s)_1 \quad\quad = \quad A\,[1/R_1 - 1/R_2]$$

$$= \quad A(R_2 - R_1)\,/\,(R_1R_2)$$

Here $A = (6.672E{-}08)(5.977E27)/c^2 = \quad 0.443708$ cm.

For $R_1 = 6.387E08$ cm, the difference is $0.996E{-}12$, or about 1 part per trillion. For 144,000 seconds, the observed difference should be 143 billionths of a second slower than the clock on Earth. This is three times that calculated from the Special Relativity correction to time.

4. Rotation of the orbit of Mercury

Before Einstein, Newtonian physics showed that the orbit of Mercury was slowly rotating about the Sun in the plane of its orbit. Like a hula hoop. The calculated rate was 5600 seconds of arc per century (**CVEP**), but after accounting for all known effects, some 43 seconds of arc per century was unaccounted for. Relativity was said to explain it.

CVEP develops a formula for the relativistic correction to θ_n, the angular displacement of Mercury in time T_n, one stage of which is:

$$\theta_n = (GM_n)^{0.5}\, T_n\, /\, (R_n)^{1.5}$$

R_n is the radius of Mercury's orbit, T_n is the orbital period, and M_n is the mass of Mercury. The subscript n means G/N mechanics is assumed.

Inserting Relativity corrections gives θ_g, the value as affected by the Sun's gravity. The difference in these two values amounts to 39 s of arc/century.

> **Comment**: The correction on R_n accounts for half of the total. This is questionable since R_n is always perpendicular to the direction of motion and therefore should not be affected by that motion. If you think I am wrong, read **RSGT**, p 81, which states this explicitly. A meter stick can

be used to measure the length of the radius of a rotating disc correctly, but will give the wrong answer if applied to the circumference.

If the correction is valid, then a similar correction should be applied in the M/M experiment to the path length perpendicular to the direction of travel. No difference was found, so either a) the theory is wrong or b) there really is no difference.

> **Comment**: Some argue that the precession can be explained by a nonrelativistic approach based on the fact that our Sun is not a perfect sphere.

5. Mass/energy relationship

Perhaps the most credible support for Special Relativity is in the predicted relationship of velocity and mass. The development of the nuclear power industry shows that $E = m c^2$ is unquestionable. In Newtonian physics, the kinetic energy (KE) of a body of mass m moving at constant velocity v is:

1) $E = mv^2 / 2$

CVEP, Appendix C, shows that the relativistic mass m is given by m_r/F, where m_r is the rest mass. The derivation is based on the Conservation of Mass and Momentum. It looks at the inelastic collision of two particles of equal mass and equal but opposite velocities so that the two particles stick together.

> **Comment**: The equations are developed using the Relativistic addition formula for two velocities, so it is not surprising that the factor F should emerge in the final result. In effect, the derivation assumes the answer.

> **Comment**: It is not clear in what direction the observer is observing the collision: is his line of sight perpendicular to the direction of the colliding particles? Or isn't this significant?

By an algebraic manipulation, the factor 1/F is expressed as the familiar infinite series:

2) $m = m_r/F$

$$= m_r [1 + (1/2)(v/c)^2 + (3/8)(v/c)^4 + \ldots]$$

For low v, the terms involving the fourth and higher powers of velocity are negligibly small and we get as a good approximation:

3) $m = m_r + (1/2)(m_r v^2)/c^2$

4) $mc^2 = m_r c^2 + KE$

The relativistic KE is thus the KE of the rest mass plus the term $m_r c^2$, which must also be an energy. Alternatively, it is as if the inertial mass of the body has increased by an increment of mass m':

5) $m' = KE/c^2$

It would seem that $E = m_r c^2$ is an incidental result, although perhaps more important to humans than all the rest of the predictions. It is empirically true even if the Lorentz Contraction is false.

Thus $m_r c^2$ is the energy equivalent of the rest mass m_r. For v approaching c, the factor F in equation 2) gets smaller and smaller and in the limit when $v = c$ the equation predicts the mass m and its kinetic energy become infinitely large.

> **Comment**: Experiments are said to confirm that the electron mass does increase because of velocity. If it comes from observing the curvature of particle tracks we are again dealing with accelerations and inferences from light beams.
>
> As far as I know, Science has not accelerated any particle to the velocity of light. With infinite mass, wouldn't it attract everything in the Universe? No. Somehow the particle becomes a massless photon. Yet it is deflected by gravity. Such faith. Such willingness to don blinders to the slipperiness of the concepts.

In Chapter 9, we noted that the energy of cosmic rays may go to at least 100 GeV, or 0.1602 ergs per particle. If they are high velocity protons, we can interpret the 100 GeV as kinetic energy beyond its rest mass. By the Einstein relation:

$$m' = 0.1602/c^2 = 1.783E{-}22 \text{ g}$$

This is an apparent mass added to the rest mass. So the total mass is:

$$m' + m_r \quad = m + 1.673E{-}24 \text{ g} = 1.79973E{-}22 \text{ g}$$

The velocity is calculated from the factor F:

$$F^2 \quad = 1 - (v/c)^2 = (m_r/m)^2$$

$$= [(1.67265E{-}24 \text{ g})/(1.79973E{-}22 \text{ g})]^2$$

$$= 8.6376E{-}05$$

and $\quad$ v/c $\quad = 0.999957$

> **Question**: What is the origin of particles with such enormous energies? How are they accelerated to within 0.0043% of the velocity of light? Or is this the velocity at which they are projected into space from stars?

Our Sun emits a solar wind consisting of protons and other charged particles, but at various velocities. As the velocities of the particles approach that of light, their energies approach those of cosmic rays. At what speed does a proton become a cosmic ray and when does it become a photon?

Science says that a sufficient concentration of energy can suddenly convert to an electron-positron pair. Perhaps a star produces photons travelling at c and some of them get slowed down to become cosmic rays (high velocity protons).

> **Question**: Along with an equal number of electrons? Do cosmic rays show a positive charge? If so, some part of the Universe must be negatively charged.

Let's compare the energies of some particles. The electron and proton masses in Table 10.2 are rest masses, which are converted to energy by the Einstein relation. The photon frequencies are calculated from the Planck relation. The cosmic ray proton mass is an apparent mass based on 100 GeV. Its frequency is beyond the cosmic ray range in some references. However, **EST**, Cosmic rays, discusses energies to 1E20 eV, or 100E09 GeV; **CVEP** has a passing mention of cosmic ray protons at 1E20 eV.

Table 10.2 Particle Energies and Frequencies

	Electron	Proton	Cosmic Proton
Mass, g	9.11E–28	1.67E–24	1.80E–22
Energy			
ergs	8.19E–07	1.50E–03	1.61E–01
GeV	0.000511	0.9383	1.0 E+02
Velocity/c	0.0	0.0	0.999957
Frequency, Hz	1.24E+20	2.26E+23	2.43E+25

Comment: These energies are so far beyond those in other sources that the possibility of a misprint arises. Perhaps 100E09 GeV was intended to be 100 GeV. If not, its frequency would be 2.42E34 Hz vs the E22-E23 range in Ch. 9.

In Ch. 14 we note another relation between E and T. Solving this for T:

T $= (2/3)(E/k)$

$= (2/3)(0.1602 \text{ ergs}) / (1.381\text{E–}16 \text{ ergs-K})$

$= 774$ trillion kelvins.

Some have said the temperature of the Sun's core is 10 million K, or 20 million K, perhaps even a billion K. No one has said anything like a trillion K.

So we have a choice of believing a fantastically high temperature + Relativity, or a fantastically high velocity without Relativity. It seems we are to accept the following:

• Increasing the velocity of a particle of mass increases its kinetic energy by ΔKE.

• The particle behaves as if its mass were increased by $\Delta KE/c^2$.

153

6. Fizeau's experiment

Armande Fizeau about 1850 studied the velocity of light in flowing transparent fluids. **STR** says he concluded that a) the ether was carried along with the fluid and b) the velocity of light relative to the tube in which the fluids were flowing was increased but not by the full velocity of the fluid. He gave an empirical formula:

1) $v_a = v_1 + v_2(1 - 1/n^2)$, where

v_1	=	the velocity of light in a stationary fluid
v_2	=	the velocity of the moving fluid
v_a	=	the velocity of light in the moving fluid relative to a stationary point, and
n	=	c/v_1, the index of refraction of the fluid.

RSGT claims that Fizeau's result agrees with the Relativistic addition formula for two velocities.

> **Comment**: Fizeau's observations were possible because he used an interference technique. But let's look at some numbers.
>
> | | | |
> |---|---|---|
> | c | = | 29,979,245,800 cm/s in a vacuum |
> | n | = | 1.333 for water at 20°C and the sodium D wavelength. |
> | v_1 | = | 22,490,056,860 cm/s in water |
> | v_2 | = | 30 ft/s = 914.4 cm/s (assumed maximum). |
>
> By G/N laws, v_a should have been $v_1 + 914.4$ cm/s. The fluid motion apparently reduced the contribution of v_2 by 514.6 cm/s, or 23 ppb.
>
> Temperature affects n by about 0.000125/°C, so 1°C changes v_1 by 2,811,257 cm/s. If the temperature of the fluid fell by only 0.00018°C, v_1 would have been smaller by 514.6 cm/s. Granted that he did detect a deviation from additivity of velocities, but how could it be attributed solely to the effect of fluid velocity?
>
> **Comment**: The index of refraction shows how much the path of a light ray is bent on passing from fluid of one density to another of different density. It is "explained" as a change in the velocity of light, being slower in a denser fluid. Are we sure fluid flow doesn't affect n? If we eliminate n from Fizeau's formula, we see that the correction to the additivity formula is some kind of interaction between the velocity of the fluid and the velocity of light in the fluid when stationary.

In the M/M experiment, the light beams passed through a single fluid (air). Instead of the fluid moving, the Earth (the tube) moved. But where is the index of refraction of air in the velocity addition formula? It is supposed to be general for any two velocities, not just light. The situation must be different from Fizeau, for how can there be any interaction between the velocity of Earth and the velocity of light?

In summary, four of six experimental confirmations of Relativity discussed here involve minute measurements, and one is suspect. The one that is unquestionably true, $E = mc^2$, doesn't necessarily validate the Lorentz Contraction.

If the Contraction is suspect, we go back to the M/M experiment. Exactly what was supposed to contract due to velocity: the apparatus, the wave length of light or space? There was no discussion of redshifts, so somehow either the nothing that is space is supposed to contract, or else the apparatus contracted in the direction of the velocity.

> **Comment:** Recall the speculations about dark matter all around us. It might be a medium.

Also, when the light beams struck mirrors, their photons were slowed to zero and reaccelerated to c to make a 180° change in direction. Were the collisions with the atoms in the mirrors perfectly elastic? How were these effects accounted for in the derivation? In any case, there is no satisfactory explanation of why the Earth's orbital velocity, or any velocity, does not affect the travel time of light.

> **Question:** Isn't there a misconception here? Additivity of velocities does not mean the velocity of light is changed any more than the velocity of a passenger walking in a train is changed if the train moves. The photons continue to chug along at a large, but finite, constant velocity.

Exhibit 10.1
Lorentz Contraction (LC)

<u>Nature of the problem</u>

In Galilean-Newtonian (GN) mechanics, velocities in a given direction are directly additive. In the Michelson-Morley experiment, the velocity of light and the velocity of the Earth were not. Why?

In the past, a "stationary" Observer A had no reason to think velocity had any effect on his measurements of "stationary" objects. Science now teaches there <u>are</u> no stationary objects and that lengths and times in a moving system are observed to be different from their values if that system were stationary.

"Stationary objects" are thus objects with zero velocity relative to the observer. Physical values for such objects are "rest" values that include any effects due to the Earth's total velocity.

<u>Reference systems</u>

The **RSGT** derivation begins directly with equations for the propagation of light beams. These require reference points in a moving system and a stationary system. Later we find statements like "as judged from the stationary system" and "take a snapshot of the moving system". So we also need observers in each system viewing the propagation in the other's system. **RSGT** invokes a 3-D coordinate system, but immediately simplifies by restricting motion to one direction.

To be less abstract, let us talk of a straight section of track on which is a railcar. On it is mounted a flashlight pointed in the forward direction of travel. As a reference point from which to measure distance, choose a fixed arbitrary point O_s on the track. Imagine a straight line extending from O_s down the track. Call this the s-axis (for stationary axis) to make our version of the equations distinct from those of **RSGT**. We also exclude movement in the y or z directions. This is the stationary System S. **RSGT** refers to this as one-dimensional travel along the x-axis from O.

Choose an arbitrary fixed point O_m on the railcar as the zero point for measurements made from the railcar as moving System M. The m-axis (for moving axis) will extend through O_m parallel to the track and hence the s-axis. The m-axis is thus slightly above the s-axis, but we can ignore this.

RSGT uses primes for its moving system (O' is the origin and x' are distances) and makes the x'-axis collinear with its x-axis.

> **Comment**: The characters O and 0 look alike. The first is the letter of the alphabet and is rounded; the second is the numeral zero and is oval. As a subscript 0 often designates initial value.

To be completely general, we put O_m initially at an arbitrary distance s_m from O_s. Distances measured from O_s are in the stationary system and are given the letter s, while those from O_m have the letter m.

Next we put Observer A at an arbitrary fixed distance s_0 measured from O_s in the forward direction; s_0 in general is different from s_m. It does not matter if s_0 is greater than or less than s_m. **RSGT** has A at O_s. We put Observer B anywhere on the railcar.

Nature of the problem

A and B can each make observations relative to O_s or to O_m concerning:

- Their initial locations
- The duration of their observations in their own Systems
- The duration of their observations in the other's System
- The velocity of the railcar
- The total velocity of the light beam.

The list includes five kinds of measurements: distance, length, time, velocity, and combinations of velocities. A can measure the following items because they are all in his own System S with no relative velocities:

- his own location relative to O_s
- time intervals

We are now in a position to state our objective.

> **Objective:** Observer A in System S observes motion in System M. Suspecting that velocity has some effect on his measurements, he wishes to develop equations which tell him how to convert his apparent measurements of distance and time intervals to "rest" values he would have obtained had he made them in System M, i.e. in a "stationary" system.

We next ask how does A determine v from a distance? It is postulated that for an external observer any object moving at constant velocity in a straight line is (apparently?) shortened in its dimensions that are in the direction of the velocity.

But what about distances? Suppose one end of a meter stick is 10 meters from a reference point that is also moving in the same direction at the same velocity. Does velocity not only shorten the meter stick in the direction of travel but also the space behind it back to the reference point?

Measurements

A can accept a distance measurement to the moving car as perceived if it is perpendicular to the direction of v and hence, by hypothesis, not affected by it. However, even though he knows the length of the railcar from observations made when it was stationary, he does not know its actual length when it is in motion parallel to its length.

For velocity, A must devise some method of observing the location of the railcar at two different times. One way might be to choose two conveniently located stationary markers a known distance apart on the track and clock the travel time between those markers. Then he can calculate the velocity.

The time interval A measures is a true time interval in the stationary system, but it may be different from the interval as measured by B on the railcar.

> **Comment:** The Hubble redshift is said to be an expansion of space at a rate proportional to the distance from us. Does this offset the Lorentz Contraction?

Let us identify the variables involved.

- c_s The velocity of light in stationary System S
- c_m The velocity of light in moving System M

As for subscripts:

- a   An observation made by A
- b   An observation made by B

In G/N mechanics, a velocity v has no effect on these measurements, so $(c_s)_a$ and $(c_m)_b$ are equal. Also:

- $(v_s)_a$ = 0   To A, System S is stationary.
- $(v_m)_b$ = 0   To B, System M is stationary.
- $(v_m)_a$ = v   To A, System M is moving forward at velocity v.
- $(v_s)_b$ = −v   To B, System S is receding in the negative direction at velocity v.

O_m is displaced from O_s in the forward s direction by the distance s_m. **RSGT** shows a picture with O and O′ displaced, but in the middle of its derivation considers only the moment when they are coincident. We will track the location of a particular photon in the light beam. At an arbitrary time t:

s   Location of the photon relative to the stationary origin O_s
m   Location of the photon relative to the moving origin O_m

Time intervals

Let us begin observations at some arbitrary time $t = t_0$ rather than at t = 0 as in **RSGT**. We have:

s_0 Initial location of the photon relative to O_s

m_0 Initial location of the photon relative to O_m.

Let any time after t_0 be t_a as measured by A in S and the corresponding time t_b as measured by B in M.

Question: What does "corresponding time" mean? We can certainly arrange to start clocks in Systems S and M simultaneously at some arbitrary instant. One interpretation is the interval from t_0 to t_b in M has exactly the same meaning as the interval from t_0 to t_a in S. But we are not to assume these intervals are of equal length.

Question: "Corresponding time" has other difficulties. It requires that we accept as true the time intervals we measure in System S. Is this correct? All of our measurements are subject to the effect of the Earth's rotation on its axis, its orbital velocity around our Sun, the velocity of the Solar System within our galaxy, and the velocity of our galaxy. Are we to assume that none of these velocities, which are in different directions, affect our measurements? Or at least that they affect all of them to the same degree?

Procedure

Let the flashlight be turned on to start a stream of photons. Let railcar be set in motion at constant velocity v at time $t = (t_0)_a = (t_0)_b$. Given that the velocity of light from a stationary source is c cm/s in System S, by G/N laws the velocity of a photon in the light beam is figured as $(c + v)$ relative to O_s and $(c - v)$ relative to O_m.

Derivation

A. Basic equations

We are to find the relations that must hold between s and m and t_a and t_b so that the velocity of light in S as seen from M is also c. The derivation assumes c is constant and finds what relations result.

For A, the distance traveled by the photon is given by equation 1).

1) $s - s_0 = (c_s)_a(t_a - t_0)$

This looks simple. It says that the location of the photon at any time t_a after the start of observations is given by the initial location, s_0, plus the product of the velocity (in this case, c) and the length of the time interval.

> **Illustration:** As a numerical example, suppose the railcar moves at velocity $v = 60$ mph at $t_0 = 12:00$ noon. Let s_0 be 1 mile from O_s. The velocity of light is 186,282 miles/second. At $t =$ Noon + 0.054 milliseconds, the photon will have travelled 10 miles down the track. The railcar will have moved a distance $(s - s_0)$ of 0.057 inches. So $s = 1$ mile + 0.057 inches.

We stress that the variables s and m in Equations 1) and 3) have been assigned to a light beam and not to the railcar. By algebra:

2) $(s - s_0) - (c_s)_a (t_a - t_0) = 0$

Next we write a similar equation observing the light beam relative to O_m. **RSGT** uses c in both 1) and 3) because of its hypothesis that c is not affected by the system velocity. We subscript c because we can easily make the two values equal later.

3) $(m - m_0) - (c_s)_b (t_b - t_0) = 0$

> **Comment:** Setting the variables with subscript 0 equal to 0 yields the **RSGT** equations.

Dividing 3) by 2), on the right side we have 0/0.

> **Comment:** This step is sure to be puzzling. We were taught that multiplying any number by zero yields zero because multiplication is simply repeated addition, so if we keep adding zeros we still have zero:

4) $(a)(0) = 0$

Now if we had written $(a)(b) = c$, we would have no difficulty accepting $a = c/b$. We find that sometimes we can also go from 4) to:

5) $a \quad = 0/0.$

This says if we divide zero by zero, sometimes there is a finite answer that may be found by special means (like calculus).

> **Example:** One of the trigonometric relations is the sine of an angle. What is the value of $(\sin x)/x$ for $x = 0$? It is 0/0. It happens that $\sin x$ can be represented by an infinite series:
>
> $$\sin x \quad = x - x^3/3! + x^5/5! - \ldots$$
>
> $$(\sin x)/x = 1 - x^2/3! + x^4/5! - \ldots$$
>
> $$= 1 \text{ when } x = 0.$$

Here the derivation assumes there is an answer that is designated as λ, lambda.

> **Comment:** Lambda is described as an unspecified constant. Not only must we assume there might be conditions under which c is <u>constant</u> for all systems moving at <u>constant</u> relative velocity, but we must also assume a relation between their equations of motion that is single, linear, and constant.

Charles H. Peterson

Comment: It seems to create a relationship from zero, i.e., out of nothing. Religion has its God doing the same thing. However, one answer is to go ahead but test the final result by observations.

6) $\lambda[(s - s_0) - (c_s)_a(t_a - t_0)] = (m - m_0) - (c_s)_b(t_b - t_0)$

Then we turn the light source around and send a light beam in exactly the opposite (negative) direction. This yields a similar equation, but the constant is not necessarily the same, so we call it mu (μ):

7) $\mu[(s - s_0) + (c_s)_a(t_a - t_0)] = (m - m_0) + (c_s)_b(t_b - t_0)$

Question: Why in the opposite direction? The critical part of the M/M experiment was the comparison of the travel times for two perpendicular paths. The equations do not address this question, or the effect of velocity itself, but only the effect of velocities in opposite directions.

Comment: The **RSGT** derivation is for a light beam proceeding along the x-axis. For events outside the x-axis **RSGT** tells us it is only necessary to add the conditions that the y and z coordinates remain unchanged. This reaffirms our understanding that it is claimed that any changes due to velocity occur only in the direction of motion. **RSGT**, p 119, also asserts that equations can be derived for System M when oriented at a general angle to System S, but gives no equations.

Then we add and subtract 6) and 7) to get:

8) $\quad m - m_0 \quad = a(s - s_0) - b(c_s)_a(t_a - t_0)$

9) $\quad (c_s)_b(t_b - t_0) = a(c_s)_a(t_a - t_0) - b(s - s_0)$

where $\quad$ a $\quad = (\lambda + \mu)/2$

and $\quad$ b $\quad = (\lambda - \mu)/2$.

These two equations tell us how the distance and time intervals perceived for a light beam in System S will be perceived in System M when corrected for the relative velocity of System M.

As a check, we divide 9) through by $(c_s)_b$ and then divide the result into 8). By 3), $(m - m_0)/(t_b - t_0)$ is $(c_s)_b$. Dividing numerator and denominator by $(t_a - t_0)$:

$$10) \quad (c_s)_b = \frac{a\left(\dfrac{s - s_0}{t_a - t_0}\right) - b(c_s)_a}{a\dfrac{(c_s)_a}{(c_s)_b} - \dfrac{b}{(c_s)_b}\left(\dfrac{s - s_0}{t_a - t_0}\right)}$$

The two quotients involving s and t_a have been identified in 1) as $(c_s)_a$.

$$11) \quad (c_s)_b = \frac{a(c_s)_a - b(c_s)_a}{a\dfrac{(c_s)_a}{(c_s)_b} - \dfrac{b}{(c_s)_b}(c_s)_a}$$

Factoring out and cancelling the factor $(a - b)$:

12) $(c_s)_b \quad = (c_s)_a \, [(c_s)_b/(c_s)_a] = (c_s)_a \, [1] = (c_s)_a$

This is what was built into the equations, namely that the velocity of light is constant and it also reconfirms that the equations were written for light beams. We will now drop the subscripts on c.

B. Evaluation of constants

From 8):

$$13) \quad s - s_0 = \left(\frac{b}{a}\right)c(t_a - t_0) + (m - m_0)\left(\frac{1}{a}\right)$$

From 9):

$$14) \quad t_a - t_0 = \left(\frac{b}{ac}\right)(s - s_0) + (t_b - t_0)\left(\frac{1}{a}\right)$$

To evaluate the constants a and b, we need two numerical conditions. One should involve the relative velocity of the two Systems; the other should

163

address the Principle of Relativity. Regarding the first condition, the **RSGT** version of 13) is x = (bc/a)t + x′. It then asserts that the origin O′ of the moving system (O_m in our version) is permanently at x′ = 0.

> **Comment:** This is confusing: x′ (our m) is the distance the light beam is from O′ (our O_m) at the time t = t′ (our t_b). It cannot simultaneously have the values x′ and 0. The only meaning x′ = 0 can have with respect to the light beam is in reference to the time when the light beam was at O′. By the equations **RSGT** offers, that is when t′ is zero. At that time, t also = 0, making x = 0. O′ does have a velocity relative to O, the origin of the stationary system, but this cannot be defined in terms of x because x is the velocity of the same light beam but as seen by the stationary observer. We need a condition on m, the position of the light beam, as related to v, the relative velocity of System O_m.

The corresponding substitution in our derivation is m = m_0. From 13):

15) $(s - s_0)/(t_a - t_0) = bc/a$.

We can also substitute in 14) $t_b = t_0$:

16) $(s - s_0)/(t_a - t_0) = ac/b$.

From 15) and 16) we get $a^2 = b^2$, or a = b.

> **Comment:** It would seem we weren't supposed to see this possibility because **RSGT** follows a different path. Its version of 15) is x/t = bc/a. Astoundingly, x/t is now identified as v, the velocity of System M as seen from System S! This is a contradiction since Equation 1) in **RSGT** has already said x/t = c.

> It is an error for x (s in our notation) to refer to both the light beam and the railcar.

The comparable result is obtained from 13) by dividing through by $(t_a - t_0)$.

17) $v = \left(\dfrac{b}{a}\right)c + (m - m_0)\dfrac{1}{a(t_a - t_0)}$

So the apparent velocity of all points s on the track as seen from System S is bc/a plus a second term.

Since s was assigned to the photon, its velocity can be found once we know what a and b are, by measurements of m and t_a relative to O_s. Setting $m = m_0$:

18) $v = (b/a)(c)$

RSGT asserts that if we calculate the velocity of another point x' (m in our notation), or the velocity of a point m relative to O_m, we would get the same expression for v.

> **Comment:** This is not obvious. Equation 17) shows this simple relation can be obtained <u>only</u> for $m = m_0$ or, curiously, for t_a = infinite time.

Let's use the letter r to subscript the variables associated with the railcar. The relative velocity of System M is then correctly expressed by:

19) $$v = \frac{(m_r)_a - m_0}{(t_r)_a - t_0}$$

> **Comment:** This says that at time $(t_r)_a$ A sees the railcar has reached $(m_r)_a$ and s_r in S while the photon has reached m_a in M. Putting in some numbers, if $(t_a - t_0)$ is 1 second, the photon has reached $m = m_0 + 186,282$ miles and the railcar has gotten to $(m_r)_a = m_0 + 88$ feet, if v is 60 mph.

The rest of the derivation appears to be meaningless. We will follow it further by assuming 18) is correct. For convenience in later use,

20a) $b/a = v/c$ by 18)

20b) $b/a = v/c - (1/ac)[(m - m_0)/(t_a - t_0)]$ by 17)

> **Comment:** Note the complication of the term involving distances from System M with times from System S.

For the second condition, the Principle of Relativity is that a given length in the moving system viewed from stationary system should appear just as long as an equal length in the stationary system viewed from the moving system. Generalizing what is done in **RSGT**, let's put one end of an object of length L at a particular point m_1 in System M and its other end at m_2. We want to know what its length appears to be in System S at any time t after t_0. From 8), we have:

165

Charles H. Peterson

21) $m_1 - m_0 = a(s_1 - s_0) - bc(t - t_0)$

22) $m_2 - m_0 = a(s_2 - s_0) - bc(t - t_0)$

> **Comment:** As has been noted, the equations were derived for the propagation of a light beam, not a meter stick. The quantities s_1 and s_2 can refer to the meter stick, but then c has to be replaced by the velocity of the meter stick, which would be the relative velocity of System M. To write these equations for the light beam, there would have to be two different values of t, namely t_1 and t_2.

However, subtracting:

23) $m_2 - m_1 = a(s_2 - s_1)$

24) $s_2 - s_1 = (m_2 - m_1)/a = L(1/a)$

Thus, L in System M appears to System S to have been multiplied by a factor 1/a. This factor will shortly be identified as the Relativity factor, and is being applied to distances, not lengths.

Now if we put a similar object of length L in System S, we can find its apparent length when seen from System M by using 8) with a substitution for $(t_a - t_0)$ from 14):

25) $m - m_0 = a(s - s_0) - \left(\dfrac{b^2}{a}\right)(s - s_0) - \left(\dfrac{bc}{a}\right)(t_b - t_0)$

26) $m - m_0 = a\left(1 - \dfrac{b^2}{a^2}\right)(s - s_0) - \left(\dfrac{bc}{a}\right)(t_b - t_0)$

In **RSGT**, t′ was chosen as 0, the starting time of the observations. Our equivalent choice would be $t_b = t_0$. It appears unnecessary since the term in t drops out in the next steps.

For any instant t_b in System M corresponding to the instant t_a in System S, we write Equation 26) for each of the two end points m_1 and m_2 and subtract:

27) $m_2 - m_1 = a[1 - (b^2/a^2)](s_2 - s_1)$

By Relativity,

28) $(s_2 - s_1)$ from 24) $= (m_2 - m_1)$ from 27)

    Comment: Thus A sees the distance $(s_2 - s_1)$ equals the distance $(m_2 - m_1)$ as seen by B.

29) $(L)(1/a) = a[1 - (b^2/a^2)](L)$

30) $1/a^2 = 1 - (b^2/a^2) = 1 - v^2/c^2$, by 20a).

However, v and b/a are given by the more complicated expressions in 17) and 20b), respectively, and really by 19) for v. For convenience:

31) $1/a^2 = F^2$.

    Comment: F is a number less than 1 since v^2 cannot be negative. By 24), the length L in System M appears shorter in System S by the factor F. To continue as in **RGST**, from 20a) and 31),

32)   b $= av/c = v/cF$

    Comment: Note that a in 31) and b in 32) depend on the relative velocity v, although **RSGT** calls them constants.

Substituting in 8) and 9),

33) $m - m_0 = \left(\dfrac{1}{F}\right)[(s - s_0) - v(t_a - t_0)]]$

34) $t_b - t_0 = \left(\dfrac{1}{F}\right)[(t_a - t_0) - \left(\dfrac{v}{c^2}\right)(s - s_0)]$

Setting the variables subscripted with 0 to 0 yields the same equations as **RSGT**.

C. Time intervals

Next we look at the effect of velocity on time intervals for any location s. From 33):

35) $s - s_0 = F(m - m_0) + v(t_a - t_0)$

> **Comment:** The distance $(m - m_0)$ is thus shortened by the factor F and then increased by the distance the railcar has travelled in the time interval $(t - t_0)$ to obtain what it looks like in System S. Only at $t_a = t_0$ is it comparable to 23).

Substituting 35) in 34) at time $t = (t_b)_1$:

36) $t_{b1} - t_0 = \left(\dfrac{1}{F}\right)\{(t_{a1} - t_0) - \left(\dfrac{v}{c^2}\right)[F(m_1 - m_0) + v(t_{a1} - t_0)]\}$

This reduces to:

37) $t_{b1} - t_0 = F[t_{a1} - t_0] - \left(\dfrac{v}{c^2}\right)(m_1 - m_0)$

Similarly for time $t = (t_b)_2$:

38) $t_{b2} - t_0 = F[t_{a2} - t_0] - \left(\dfrac{v}{c^2}\right)(m_2 - m_0)$

Subtracting:

39) $(t_b)_2 - (t_b)_1 = F[(t_a)_2 - (t_a)_1)] - (v/c^2)(m_2 - m_1)$, from which

40) $(t_a)_2 - (t_a)_1 = (1/F)[(t_b)_2 - (t_b)_1] + (v/Fc^2)(m_2 - m_1)$

So the time interval in System M appears longer in System S by the factor 1/F and plus an amount depending on distance involved and the velocity of System M.

> **Comment:** This result is different from that in 23). We can talk about the location of two points on a meter stick at any time t. But to talk about two different times means two different locations.

Looking more closely, the terms in m do not appear in **RSGT** because the time comparison was made for $m = m_0$. They vanish for the point $m = m_0$, which is the general initial point. BUT NOT FOR ANY OTHER POINT.

If we accept the peculiar algebra, we do get the LC equations. If we don't, we still have the problem of how to deal correctly with the relative velocity of the two systems. We need to relate m_r or v to s in some way. The way used in **RSGT** has no basis and violates the rules of algebra. What we need is a mechanism for how velocity affects our perception of light beams. Does it contract spatial dimensions and expand time? Or does it affect light itself somehow?

Loose ends

There are two more loose ends. Generally, when techniques such as that in 6) and 7) are used, one finds values for the parameters. In **RSGT** no further thought is accorded them. We find:

41) $\quad \lambda \quad = a + b = (c + v)/F$
$\quad \quad \mu \quad = a - b = (c - v)/F$

So the parameters are not arbitrary, and in fact are not even constants because of the presence of v. Which is another violation of the ground rules. It also reminds us the relevance of the equation involving μ is obscure. In applications, only the forward velocity is ever mentioned; the velocity in the reverse or in the perpendicular directions is not invoked.

The other loose end involves 33) and 34). In 33), v is supposed to be the relative velocity of System M. In 34), the reciprocal of the factor v/c^2 is a velocity and is clearly much greater than the velocity of light.

Comment: HEF, p 9-08, notes that any particle of mass m is thought to be associated with a wave of frequency given by:

$$E = h\nu = mc^2/F,$$

and the velocity of propagation of the wave has to be defined as $V = c^2/\nu$ in order to satisfy the wavelength relation for radiation. I suppose this means that all three quantities (v, wavelength, and frequency) can be measured independently. But unless the observed velocity v is equal to c, the defined velocity V will be greater than c. The reference goes on to say no physical meaning can be attached to a velocity greater than that of light in a vacuum. Which seems to be a loose end.

Addendum
Expression for Addition of Velocities

The following assumes the Lorentz equations apply. Suppose System M is moving in a straight line at constant velocity v_m relative to point O_s in System S.

Suppose object X in System M is moving in a straight line from an initial point m_0 and in the same direction as v_m at constant velocity v_x relative to O_m. What is v_a, the observed velocity of X relative to point O_s?

1) $v_a = (s_a - s_0)/(t_a - t_0)$

The velocity of X relative to System M, regardless of whether it is directed the same or opposite to M is:

2) $v_x = (m_b - m_0)/(t_b - t_0)$

Substituting the Lorentz equivalents from 33) and 34) in the main text for the terms involving m_b and t_b:

3) $$\frac{(s_a - s_0) - v_m(t_a - t_0)}{F} = \frac{v_x}{F}[(t_a - t_0) - \left(\frac{v_m}{c^2}\right)(s_a - x_0)]$$

The term v_m takes the place of v as the velocity of the railcar, i.e., System M. But remember that these equivalents are invalid since v was introduced improperly into 8) and 9) to get 33) and 34) in the main derivation.

Multiplying by F and then dividing through by $(t_a - t_0)$ in 3):

4) $$\frac{s_a - s_0}{t_a - t_0} - v_m = v_x[1 - \frac{v_m}{c^2}\left(\frac{s_a - s_0}{t_a - t_0}\right)]$$

Substituting from 1):

5) $$v_a - v_m = v_x[1 - \frac{v_m}{c^2}v_a]$$

6) $\quad v_a = \dfrac{v_m + v_x}{1 + \dfrac{v_m v_x}{c^2}}$

So we have three applications.

- If either v_m or $v_x = c$, then $v_a = c$.

- If $v_m = v_x = c$, $v_a = 2c/2 = c$.

This is for one flashlight travelling at c while sending off photons at c.

- For the case of two base-to-base flashlights, we use $-v_x$ in 6) to get:

$v_a = (c - v_x)/[1 - v_x/c] = c$

Exhibit 10.2
Gravitational Redshift

System description

Consider a disc of radius R rotating at constant angular velocity w (omega) about an axis perpendicular to the disc at its center O. On the disc are two identical clocks, one at O and one at its edge. There is also a stationary observer at O. The tangential velocity at the edge of the disc is $v = \omega R$.

Time and velocity

In Special Relativity, a second of time in a system moving in a straight line is seen by a stationary observer to be longer by the Relativity factor 1/F. This is generalized to any time interval T:

$T_s = T_m/F,$

where s means stationary and m means moving. So if T_m is one second, it will be seen as T_s, a longer interval, in the stationary system. Note the "seen". It implies T_m is not actually lengthened.

Time and gravity

General Relativity asserts that whatever effect is observed from a centripetal (center directed) acceleration will be reversed for a centrifugal (repulsive) accleration. **CVEP** asserts that working out the length and time changes due to a gravitational field is very difficult. The results are for relatively weak fields such as that of our Sun. So we really aren't told the basis for the assertion as to the effect of a gravitational field on time.

Let's assume the assertion that T_m is lengthened in a gravitational field and shortened in a centrifugal (repulsive) acceleration is true.

Time and rotating motion

The clock at the edge of the disc is subject to a centrifugal acceleration due to the rotation of the disc, **RSGT** asserts that the acceleration shortens T_m.

$$T_s = F\,T_m$$

> **Comment:** The factor F is due to the <u>linear velocity</u> at the edge of the disc, because Special Relativity was derived for motion in a straight line and not in the presence of a gravitational field. So we are not off to a good start.

Frequency and velocity

Frequency is the reciprocal of a time interval, so the frequency of ticks of the clocks on the rotating disk are related by:

$$\nu_m = F\nu_s,$$

where ν_s is the number of ticks/second of the clock as determined by a stationary observer when the disk is at rest and ν_m is the number of ticks/second of the clock as determined by the stationary observer when the disk is rotating at ω revolutions/second. F is given by an infinite series by the binomial theorem:

$$\nu_m = \nu_s\,(1 - v^2/2c^2 - \dots)$$

Substituting the equivalent angular velocity:

$$\nu_m = \nu_s\,[1 - (1/c^2)(\omega^2 R^2/2)]$$

$$= \nu_s\,[1 - (1/c^2)(B)], \text{ where } B = \omega^2 R^2/2.$$

So far, this is only Relativity and algebra.

Potential

The electrostatic potential (in volts) at a point is the electric energy at that point divided by the charge at the point. The factor B has the units of a velocity squared, which is an energy per unit mass:

$$\text{Joules/kg} = \text{(newton-m)/kg}$$

$$= (\text{kg-m/s}^2)(\text{m}) / \text{kg} = (\text{m/s})^2$$

B is interpreted as a gravitational potential, as a counterpart to volts.

Frequency and work

Now work in general is:

$$W = \text{Force x distance x} \cos\theta = Fs \cos\theta,$$

where θ (theta) = the angle between the direction of F and that of s.

The positive direction for R is out from O. In discussing the gravitational pull on the clock, both F and s are directed toward O so θ is 0, $\cos\theta$ is +1, and the work is positive. For centrifugal force, both F and s are directed away from O, so again the work is positive. **RSGT** notes that if the work is done against the centrifugal force to move the clock from the edge to O, this work is negative. The derivation continues:

$$v_m = v_s [1 + \text{ø}/c^2], \text{ where ø} = -B$$

Comment: Phi is the gravitational potential at R. There is no name for the unit of gravitational potential comparable to volts in electrostatics.

Frequency and gravity

Then this is generalized to gravitational fields for which $\text{ø} = -GM/R$:

$$(v_s - v_m)/v_s = (GM/c^2) / R = A/R, \text{ where}$$

G = the universal gravitational constant
M = the mass of the body creating the field
R = the radius of the body

    c   = the velocity of light
    v_m = the shifted frequency and
    A   = GM/c^2. The units of A are meters.

A cannot be negative, so the frequency of a photon emitted from an atom on the surface of a rotating body is always less than that for an atom at the center of the body.

Chapter 11. Technology: Structure of Matter

Molecular Models

Empedocles, about 460 B.C., taught that matter was made of four elements: water, air, fire, and earth. There are millions of very small particles of each of these four substances, permanent and unchanging. We sense things because particles leave an object and fall on the "pores" of our senses (**TSC**).

> **Comment:** This should include the idea that sunlight has a particle nature.

He also taught that change is purposeless, but every change has a cause and is governed by Chance and Necessity. Earlier, these beings ruled even Zeus (**HWP**).

> **Note:** The first change, appearance of the Universe, therefore had a Cause. But where did Chance and Necessity come from?

Anaxagoras (500?-428 B.C.) held that matter is <u>infinitely</u> divisible (**HWP**). Things don't originate by necessity and chance (**HWP**), but are caused by a world mind (**BTGP**).

Democritus (c460-357 B.C.) taught that everything is made of solid, incompressible atoms, too small to be visible and <u>indivisible</u>, infinite in number, and in motion always. There had to be space between them or motion would be impossible. The universe has no purpose; there are only atoms governed by mechanical laws. Nothing happens just by chance (**HWP**).

Proposition I of Book X of the Elements by Euclid (c300 B.C.) states that if some quantity A is greater than B, it is always possible to remove a part of A to leave a quantity smaller than B (**TMEW**, p 262). This conflicts with Democritus, but agrees with Anaxagoras.

These few examples call attention to the centuries-old human quest for answers in which there were glimmerings of what Science now teaches as "truth".

> **Comment:** Why are the ideas of Democritus on divisibility of matter given precedence over those of Anaxagoras? Purposelessness may be

stressed since it fits nicely into the secular view of the Universe. Democritus holds out for mechanical laws but does not explain their origin. Are there other writers we don't hear about?

Comment: They drew Laws of Nature from Nature. Rain fell to Earth and eventually flowed via rivers into the ocean. No purpose. Really? The effect (purpose?) of the hydrological cycle is to make soil and put drinking water all over the Earth.

Roger Bacon in the 13th century declared that we can investigate by argument or by experiment. The former is insufficient, so the only (?) way to explain our observations is to rely on experiments. However, he also is denying the validity of argument and intuition.

Three centuries later Francis Bacon rediscovered Democritus from studies on the compressibility of air that showed matter must be made of particles separated by empty space. After Robert Boyle and others, it became accepted that a gas was made up of an enormous number of particles called molecules travelling in random directions at high velocities and constantly colliding with one another. It was assumed that (somehow) the molecules did not lose energy on collision, perhaps by being perfectly elastic.

Atomic Models

A. Planetary models: descriptive

After 2000 years, Science finally took one step beyond the Greeks. The proposed model was still quite simple, but required many decades of work. Many writers have described the fascinating experimental work involved so we will take note of only certain significant steps.

1. Spectral lines

Newton (1642-1727) showed that sunlight was made up of a continuous series of colors. Since light cast sharp shadows, it had to be made up of corpuscles (particles). Thomas Young (1773-1829) showed by interference experiments that it had a wave nature, and made the first measurement of its wavelength. The colors in sunlight thus differed in wavelengths and frequencies. In 1873 Maxwell deduced that light was part of an electromagnetic spectrum.

It was known that light from incandescent solids after passing through a slit, a focusing lens and a prism or a fine grating produced a continuous spectrum made up of closely spaced lines. Assuming the lines corresponded to different wavelengths and hence frequencies, something must be oscillating or vibrating within the atom. E. J. Balmer in 1885 deduced a single formula for all wavelengths, λ (lambda), in meters of the lines in the visible light range from hydrogen:

$$\frac{1}{\lambda} = R\left(\frac{1}{2^2} - \frac{1}{n^2}\right)$$

where R = Rydberg Constant, 1.0971E07/m
        n = 3,4,5, . . .∞

It was empirical, meaning it was not derived from any theoretical basis. It merely fitted the data. The constant was evaluated from known wavelengths.

2. Photons

In 1900 Max Planck proposed that radiant energy was transmitted in discrete amounts he called quanta that behaved like particles. But even more innovatively he assumed that the energy of a quantum was directly proportional to its frequency: $E = h\nu$, where h is Planck's constant and ν (nu) is the frequency in Hertz.

Combining Einstein's $E = mc^2$ with Planck's Law led to the inference that photons were associated with mass:

$m = E/c^2 = (h/c^2)\,\nu$

The relativistic mass m is given by:

$m = m_0 / F$, where $F = [1 - (v^2/c^2)]^{0.5}$.

Although the rest mass, m_0, is assumed to be zero for photons, they travel at the velocity of light. This makes F zero, and m indeterminate, but possibly different from zero.

Question: What determines how much mass the photon manifests? Its velocity is constant at c, so (somehow) it must be the frequency. Let's compare some masses to that of an electron, m_e.

$m = [(6.6262\text{E}{-}27 \text{ erg-s})/(2.9979\text{E}10 \text{ cm/s})^2]\,\nu$

$= (7.373\text{E}{-}48 \text{ g-s})(\,\nu \text{ per s})$

Charles H. Peterson

Table 11.1 Calculated Photon Masses

Ray	v, Hz	Mass, g	Mass/m_e
Yellow light	5E14	3.69E–33	0.000004
Gamma	1E20	7.37E–28	0.81
Cosmic	3E23	2.21E–24	2430

While cosmic rays are not electromagnetic radiation, this illustrates the enormous range of masses associated with photons and particles.

About 1924, Arthur H. Compton bombarded metals and gases with X-rays. Most of the radiation passed right through the targets, but some was scattered in all directions. Analysis of the results indicated that the incident radiation was deflected as if it were solid particles, like billiard balls, supporting these ideas:

- Photons have momentum, so photons must have mass.

- High energy photon interactions with electrons obey the Conservation Laws of Energy and Momentum.

- There is something orbiting the nucleus that behaves like particles (electrons).

Comment: With the later acceptance of the wave model of the atom, this explanation of photon deflection should be discarded. If the photon is a particle, how can it be reflected off a wave?

3. Rutherford model

In 1911, Ernest Rutherford postulated that an atom had a central positively charged nucleus and negatively charged electrons orbiting it in circular paths. He inferred that atoms are mostly empty space because, on bombarding them with alpha particles from radium, most of these particles were not deflected. A major objection was that an electron orbiting in the electromagnetic field around the proton should emit radiation, thus losing energy. The atom should collapse. Also, atoms emitted energy under certain conditions and yet remained as atoms. No one could explain this.

4. Bohr model

In 1913, Niels Bohr "solved" the problem of no energy loss from orbital electrons by assuming their angular momentum was quantized, i.e., could only have values that were integral multiples of $h/2\pi$.

An older text (**UP**) uses the de Broglie concept of matter waves (Wave Models, below): an electron may be a standing wave around the nucleus, requiring that the circumference of the circle whose radius is r_n that it travels in is an integral number of wavelengths:

$2\pi r_n = n\lambda = nh/(m_0 v)$, from which

$m_0 v r_n = n (h/2\pi)$

The quantity $m_0 v r_n$ is the angular momentum. De Broglie came later, so Bohr's proposal was a pure conjecture.

Combining his assumption with Newton's and Coulomb's laws yielded a formula for the energy in each orbit.

$$E_n = -\frac{1}{n^2}\left[2m_0\left(\frac{\pi A}{h}\right)^2\right] = -\frac{1}{n^2}[E_1]$$

$A = Q_p Q_e / (4\pi\varepsilon_0)$

$\quad = (1.602189E-19 \text{ C})^2 / 4\pi[8.854188E-12 \text{ F/m}]$

$\quad = 2.307113E-28 \text{ J-m} = 2.307113E-19 \text{ erg-cm}$

E_1 is the energy of the ground state. Substituting numerical values for m_0 and h:

$E_1 \quad = -2.17990E-18 \text{ joules} = -13.6 \text{ eV}.$

$E_n \quad = (1/n^2)(-E_1)$

$\quad = (1/n^2)(-2.17990E-18 \text{ joules})$

Explanation: E_n is the energy associated with orbit n, whose radius is r_n. The minus sign results from the convention that the potential energy of the electron is zero for r_n = infinity. N is the principle quantum

179

number and runs from 1 to infinity. It denotes successive orbits counting out from the nucleus. The ground state is the orbit with the lowest energy (n = 1). It does not radiate energy.

The formula permits calculation of the energy change in moving an electron between any two orbits. Amazingly, the wavelengths corresponding to these transitions do match the observed spectral lines with a high degree of accuracy.

Question: Another puzzle: what forces the atom to give up the extra energy? Or is it <u>pulled</u> out of the atom by some undetected force? We may speculate in this manner since many are searching for a mysterious "dark matter".

Heat will be radiated or convected from a hot body to a cooler one continuously until temperatures have equalized. The hot body apparently has no choice: it <u>must</u> surrender the heat.

Comment: Let those who would draw Laws of Human Behavior from the Laws of Nature ponder this well.

5. Rydberg update

Combining the orbital energy formula with the Planck equation yields the Rydberg constant:

$$R = [\frac{2\pi^2 m_0}{ch^3}]A^2 = [B]A^2$$

B = $2\pi^2$ (9.10953E–31 kg)/(2.997924E08 m/s)(6.62618E–34 J-s)^3

= 2.061652E62/joules^2-meter^3

R = 1.097370E07/m, vs the empirical
    1.096778E07/m.

CVEP notes that even the small discrepancy could be explained. The derivation was based on considering the proton as having infinite mass, whereas the center of rotation of the proton-electron system was actually between them. Dividing the calculated R by the factor (1 + m/m$_p$), or 1.00054462, yields 1.096773E07. This is low by 5 ppm. The Relativity correction on mass would make the calculated value high by 22 ppm.

Comment: Astounding! How would those who would discard G/N mechanics explain this success? But does this prove the relationship is true? No. Consider the logic:

- We observe some phenomenon.
- We develop a hypothesis to explain it.
- We predict some effect by our hypothesis.
- We do experiments and make observations.
- If observations agree with predictions, we say this is proof.

Science can really only claim that the hypothesis is <u>consistent</u> with the phenomenon. How do we know it was seen accurately? Recall our examples of common sense? And then Relativity tells us that what Newton saw was only a special case of a more general explanation.

Comment: The model explained nothing. It merely replaced the unexplained mystery with another unexplained mystery that was to be accepted on faith that it was always true. Until the next model. What is the final authority? Our ability to test, to observe, or to think?

Comment: Despite the violation of Maxwell's Law, Science chose to accept the Bohr postulate. This is a form of the syncretism that some try to promulgate in Religion. How is it that Scientists, who worship Logic, have no difficulty believing that conflicting explanations for a given phenomenon are equally valid? Isn't this is a dangerous concept to disseminate in our society? It undermines the logic of Logic itself.

B. Planetary models: size

1. Diameter estimates

It is natural to want some idea of the physical dimensions involved in atoms. While mass, charge, and other properties are readily available in the technical literature, the physical sizes are not. Of the various estimates found, we will use:

- **CRC**: The diameter of the first Bohr electron orbit.
- **EST**, Atom: The diameter of the proton.
- **EST**, Fundamental constants: Electron radius.

2. Physical characteristics

Table 11.2 shows estimates of the sizes and mass densities of atomic particles. The proton mass is 1836 times that of the electron, but the amount of charge on each is exactly the same, 1.602E–19 coulombs. No difference has ever been found in the quantity of electricity in the positive and negative charges.

Table 11.2 Some Physical Characteristics of Elementary Particles

	Unit	H Atom	Proton	Electron
Diameter	cm	1.06E–08	2.80E–13	5.64E–13
Volume	cm^3	6.24E–25	1.15E–38	9.39E–38
Mass	g	1.67E–24	1.67E–24	9.11E–28
Density	g/cm^3	2.68E+00	1.45E+14	9.70E+09

Comment: The proton mass density is incredibly high, like that of a neutron star (see Table 11.3), yet it generates a very weak gravitational field.

Table 11.3 Mass Densities of Stellar Objects, g/cm^3

White dwarf	2E+05 to 1E+09
Neutron star	6E+12 to 1E+17
Black hole	Possibly > 9E+19

C. Planetary models: mechanics

1. Gravitational force

Scaling up the hydrogen nucleus to 1 millimeter in diameter, the electron diameter would be 2 mm and its orbital radius would be 18 meters. The gravitational force on it, $m_e g_p$, is 3.63E–42 dynes. For comparison, the Earth's gravitational force on a 1 gram mass on the surface of the Earth is 981 dynes.

2. Electrostatic force

The attraction is mainly the electrostatic force that by Coulomb's Law is:

$$F = A/r_n^2 = (2.307E-19 \text{ erg-cm})/(5.3E-09 \text{ cm})^2 = 8.21E-03 \text{ dynes},$$

or 2.26E39 times as great as the gravitational force.

3. Electron orbital velocity

The velocity v of the electron in the hydrogen atom can be calculated from a rearrangement of the equation for angular momentum in § A.4:

$$v = nh / 2\pi m_0 r_n$$

An expression for r_n can be obtained by equating the centripetal and electrostatic forces on the electron, and using the above expression to eliminate v:

$$r_n = n^2 (h/2\pi)^2 / (Am_0)$$

$$= n^2 (5.292E-09 \text{ cm})$$

In the smallest orbit, n = 1, and v = 2.188E08 cm/s, or c/137. Relativity effects in atoms are tiny: the correction factor F is 0.9999734 (27 ppm). The orbital circumference, C, may also be the wavelength:

$$C = 2\pi (5.292E-09) = 3.325E-08 \text{ cm},$$

and its frequency is 6.580E15 Hertz.

4. Potential energy

Potential energy (PE) and kinetic energy (KE) were described in Chapter 9. Let us apply this information to an atom. Consider an electron approaching a proton from Point 1 at an infinite distance from the proton. **CVEP** applies the Law of Conservation of Energy: for an isolated system, the total initial energy, ΣE, must equal the system's total final energy at any Point 2.

$$PE_2 + KE_2 = PE_1 + KE_1 = \Sigma E$$

The larger body is assumed to be at rest so all the energy changes can be figured as occurring with the smaller body. It is at rest so its KE is zero. By

183

convention, its PE is also zero, and hence the total energy of it, and of the system of two bodies, is zero.

Comment: We need a rationale for how the smaller body ever starts moving toward the larger one, since the attractive force at infinity is zero. Science says that no matter how far away the electron is, it will still experience a tiny attractive force. Religion has a similar concept: no matter how far we run from God, He is still there.

Inserting the previously found expressions:

$$\Sigma E = -A/R + \frac{1}{2} m_0 v^2 = 0 + 0 = 0$$

Using the values for r_n and v found in Section C.3,

$$PE_2 = -A/(5.292\text{–}09 \text{ cm}) = -4.36E\text{–}11 \text{ ergs}$$

$$KE_2 = (9.1095E\text{–}28 \text{ g})(2.188E08 \text{ cm/s})^2/2 \\ = 2.180E\text{–}11 \text{ ergs}$$

Half the PE lost is retained as KE of the electron; the rest is radiated as photons.

Comment: The orbital radius is 37,800 times the proton radius. Our Moon distance is only 60 times the Earth's radius.

To examine conditions at a Point 3 closer to the proton, we can't use the Bohr equation for v because the n can't be less than 1. Instead, we equate the centripetal and electrostatic forces:

$$m_0 v^2/R_3 = A/R_3^2, \text{ from which}$$

$$R_3 = A/(m_0 v^2)$$

Let's consider R_3 where $v = 0.31623$ c.

$$m_0 v^2 = (9.1095E\text{–}28)(9.4803E09)^2 \\ = 8.187E\text{–}08 \text{ ergs}$$

$$R_3 = (2.3071E\text{–}19)/(m_0 v^2) = 2.818E\text{–}12 \text{ cm, or} \\ 20.1 \text{ x proton radius.}$$

$$KE = (1/2)m_0 v^2 = 4.093E\text{–}08 \text{ ergs}$$

$$PE = A/R_3 \qquad = -8.186E\text{–}08 \text{ ergs}$$

Electrostatic force $= A/R^2 = 29053$ dynes

Centripetal force $= m_0v^2/R = 29052$ dynes.

These results are similar to those for the first Bohr orbit, although the forces are more than 3 million times as large. So why doesn't the electron continue on to the nucleus? Either it cannot acquire a velocity greater than c/137, or there is a repulsive force from the nucleus. The Relativity correction factor is now 0.9487, so on closer approach, it should be considered.

Before we move on, let's look further at the kinetic energy of the electron in the first orbit:

$$KE = 2.180E{-}11 \text{ ergs}$$
$$= (2.180E{-}19 \text{ joules})/(4.1868 \text{ joules/calorie})$$
$$= 5.207E{-}19 \text{ calories}.$$

This seems very small, but it means an energy density of 572 million calories per gram. A water molecule subject to such thermal energy would instantly break up into bare hydrogen and oxygen nuclei.

There is an even greater mystery. The energy of the rest mass by the Einstein relation is 8.19E–07 ergs, or 38,000 times this. So if the electron is a wave, all this energy should be available.

Comment: Here is a curious coincidence. The ratio of the radius of the first orbit to the radius of the nucleus was also 38,000. Let's look into this.

$$r_n/r_p = [(1^2)(5.292E{-}09)] / (1.4E{-}13)$$
$$= 37800 \text{ (dimensionless)}$$

For the kinetic energy ratio (rest mass equivalent/ orbital KE):

$$\frac{(KE)_r}{(KE)_n} = \frac{1}{2F}\left[\frac{nch}{\pi A}\right]^2$$

$$(KE)_r/(KE)_n = (1/2) [(1)(c)(6.6262E{-}34)/\pi A]^2$$

$$= 37558 \text{ (dimensionless)}$$

Now I don't know if there is some theoretical reason why these two ratios should be equal, but they are close enough to wonder about. The difference of 242 is 6443 ppm. Recall that the orbit of Mercury precesses 5600 seconds of arc per century, and that Relativity claimed to explain the classically unexplained discrepancy of 43 seconds per century (7679 ppm). Our two ratios are in better agreement. Let's equate them.

$$r_p = 2F\left[\frac{\pi A}{nch}\right]^2 \quad r_n = F\frac{A}{2m_0c^2}$$

With F = 1, the diameter of the proton is 281.8E–15 cm, only slightly larger than the value in Table 11.2, thus supporting the equality of the two ratios.

This can be manipulated algebraically by substituting the value of A in terms of r_n from § C.3. and:

$$m_0c^2 = E_0 = h\,\nu = h\,\omega/(2\pi)$$

$$r_p = \frac{F}{2}\left(\frac{2\pi}{h\omega}\right)\left[n^2\left(\frac{h}{2\pi}\right)^2\left(\frac{1}{r_n m_0}\right)\right]$$

Simplifying and then multiplying both sides by r_n:

$$m_0r_n^2\,\omega = F[nh/(2\pi)][n/2][r_n/r_p)]$$

The angular momentum in the nth orbit is quantized as shown by the factor $nh/2\pi$, but the orbit number really comes in as the square and there is an additional factor of the ratio of the radii.

> **Comment:** This conflicts with Bohr. Science may say the correct result for the radius of the proton is fortuitous, as is the near equality of the ratios. The Relativity "explanation" of the 43 s of arc may also be fortuitous, since Science does not have a general solution for the motions of even only three bodies in space.

This is an illustration of the logic of Science in exploring the unknown. Let us go on with our critique.

> **Question:** What is the structure of an electron that permits such enormously dense concentrations of energy? What holds it together? Yet an atom can lose an electron as a discrete bundle of charge, as by

heating the atom to a very high temperature. What regulates the process whereby a quantity of energy is assembled from the energy wave around a nucleus and vomited up as a discrete electron? The process is reproducible to several decimal places. In chemical reactions, one atom can capture one or more "electrons" from another atom.

Question: How does an electron retain its identity in a copper wire as it pushes through clouds of electron stuff distributed as waves around nuclei?

<u>Wave Models</u>

However, these calculations may be meaningless because we now have a wave mechanical model. The "position" of an electron in an atom is merely the point of maximum probability at any instant.

Question: Spectroscopic evidence is said to support the concept of electronic shells in atoms. Granted that somewhere the negative charge reaches a maximum, is the rest of it spread out in all directions over only the shell it is in? And if it does, why is this so? Quantum theory says that each shell has a specific energy. What keeps the energy at these values? How thick is each shell?

CVEP shows that an algebraic manipulation of $E = mc^2$, where m is the relativistic mass of a particle, and the momentum formula, $p = mv$, leads to:

$$E^2 = p^2c^2 + m_0^2c^4$$

Then it is <u>asserted</u> that this relation is completely general and, even though it was derived for particles with mass, it applied to electromagnetic radiation. The rest mass m_0 for all such radiation is <u>assumed</u> to be zero, so we are left with:

E = cp, which, since

$hv = \lambda vp$, leads to $p = h/\lambda$.

Having produced a relation involving an assumption that applies only to radiation, Science then calmly applies it to matter. About 1924 Louis de Broglie proposed that matter could also exist in wave form whose wavelength is given by this equation. Science backtracks further and replaces p with its <u>non</u>relativistic momentum equivalent (**CVEP**, p 713):

$m_0v = h/\lambda$

or, λ = $h/(m_0v)$, the wavelength of a matter wave.

Complications

Further observations complicated Science's picture of atomic structure. Experimental support was claimed for a variety of new particles: neutrons, positrons, anti-protons, neutrinos, and antineutrinos. However, the equations describe not atoms but their <u>behavior</u>. Let me quote Bazzoni, professor of experimental physics:

> ". . .saying that. . .an atom behaves. . .as if it were one of these artificial . . . representations. . .is a very different thing from saying that the atom <u>is</u> a set of waves or that it <u>is</u> built up of a system of orbits. . .We do not <u>know</u> what an atom <u>is</u> - we only know how it acts in a few sets of circumstances. . . ."

> ". . .the mathematical physicists have either adopted certain sections of (previous) theory. . .which happened to fit the conditions disclosed, or else they arbitrarily altered previously known mathematical procedures by the introduction of new rules of calculation. . .so that the results of further calculations. . . (agree) with the observations. (This) is not surprising. The mathematics was specially selected and specially handled to secure this concordance. . . If this demonstrates anything it is something connected with the way the brain functions, not the way in which the atom functions nor with the way in which the Creator thinks." (**EAM**, p 90).

A conceptual model is a specific proposal for explaining some observation by some empirical law. But to say the Law <u>explains</u> what we observe is pretentious and absurd. It merely <u>describes</u> what we have observed. Beyond absurdity, such a practice diverts us from asking what is the reality behind the Law. It is only in this way, however, that Science can claim to have found the answers to our questions about our Universe.

Dirac in 1930 proposed that in the vacuum of space electron-positron pairs continually appear and disappear in extremely short times.

> **Comment:** Why only in space? Why not also in our own bodies? Even if some potential drives the pair into our Universe, what drives them back? This violates the unidirectional action of the Second Law of Thermodynamics.

It is conceded that this process <u>does</u> <u>not</u> <u>conserve energy</u>. Because it happens so fast it is called a virtual process. The Second Law implies a concentration of energy has a higher potential, or perhaps a lower entropy, than an electron-antielectron pair, and so it tends to form the pair.

Question: When the positron collides with an electron, the pair disappears and energy is always the result. Is this energy at a lower potential than the original energy?

Current picture of atomic structure

It is not the purpose of this book to present the full historical development of atomic theory but rather to focus on some of the main concepts on the nature of our Universe and what evidence supports them. These concepts are set forth confidently in the technical literature and presented as if there is no question about them. Let us, however, heed these words:

Matthew 5:41 "And whosoever shall compel thee to go a mile, go with him twain."

Comment: Or five. Or ten, hoping that this path will lead us somewhere. . . .

Despite all the talk about matter being made up of waves, modern efforts strive for explanations in terms of particles. The experimental work continues by accelerating particles, not waves, to high velocities and then directing them against other particles, not waves, and observing the trajectories of the resulting particles, not waves.

There seem to be many kinds of particles, four fundamental forces, and several particles that transmit those forces. The term "mediate" is used, which has a kind of religious tone to it. The mediating particles are called bosons for forces, and fermions for matter. Science describes these particles by familiar properties: mass, charge, energy, and lifespan. We need to understand another property: spin.

A. Spin

   1. Fine line structure

The lines on spectrograms are often made up of closely spaced finer lines. These were "explained" by supposing that the electron orbits in a particular shell of the Bohr atom were not all circular or of the same size, and would have slightly different energies. So when an electron dropped from any outer orbit to an inner orbit, slightly different energies would be released according to which orbits were involved. The fine lines were due to the different angular momenta. Experiments using magnetic fields then showed that the lines could be split into even finer lines. Question: to what phenomenon in the atom did these correspond?

2. Magnetic fields

Atoms like hydrogen and silver have a single electron in their outermost orbits that with the positively charged nucleus makes an electric dipole. The orbital motion of the electrons, which is a tiny electric current, produces a magnetic field by interacting with the electric field of the nucleus. The atoms behave as if they were permanent magnetic dipoles.

3. Stern-Gerlach experiment

In 1921 Otto Stern and Walther Gerlach investigated the angular momentum of atomic electrons by passing a beam of silver atoms vaporized at 1000 K through a horizontal slit, then through a non-uniform magnetic field to a vertical glass plate (**CVEP**).

The expected trace was a continuous band shaped like the intersection of two circles. The results had a surprise: the traces were two curved <u>bands</u> bounding an area shaped like the intersection of two circles. **CVEP** asserts that by the Bohr theory, if the angular momentum of the ground state were equal to h/2π, there should have been three bands, but doesn't say why.

The theory did not predict the two traces. The reasoning is not clear, but it seems the conclusion was they must have been due to something other than <u>orbital</u> angular momentum. The interpretation was that the silver atoms could take only one of two positions in the magnetic field, i.e., the angular momentum of the atom was indeed quantized.

> **Comment:** This does not seem to be the same kind of quantization as the energy of the orbits because it occurred only in a nonuniform magnetic field.

Next, **CVEP** says the two bands meant that the Bohr model was wrong in that the ground state angular momentum was zero, not h/2π. It was <u>postulated</u> that, whether in or out of an atom, an electron has an inherent spin whose magnitude is given by:

$$S = [s(s + 1)]^{0.5} (h/2\pi)$$

where s can only be 1/2 , instead of a range of values. The spin can be in either direction, so S is ± 1/2 . The actual value of spin-1/2 is:

$$J = [(1/2)(3/2)]^{0.5} (h/2\pi), \text{ or } 9.13E{-}28 \text{ erg-s}$$

The unit erg-s is identical to g-cm²/s, the unit for angular momentum. The surface velocity of the electron is 300 x the velocity of light. Another loose end.

Comment: This value is the same for all spin-1/2 particles, irrespective of their mass, size, charge, or matter/antimatter structure.

CVEP asserts that the angular momentum L of the hydrogen atom is given correctly by:

$$L = [\ell(\ell + 1)]^{0.5} [h/(2\pi)],$$

where ℓ is a new quantum number taking on the values 0, 1, 2, . . . (n–1). Each of the energy states calculated from the equation in Section A.4 under Atomic Models can have n possible values. In the ground state, $\ell = 0$ and L can only be 0.

Comment: Without a derivation from theory, we must suspect the "formula" was made to fit the facts, not that the facts prove the formula.

Comment: If the electron is a wave, how can a distributed charge that has a maximum that can wander anywhere within the atom have a defined spin? Belief in a point that spins as an explanation for a force is like an article of faith in a religion.

Comment: The main lines on a spectrogram were "explained" as a physical change: an electron emitting or absorbing energy while changing orbits. To what physical change do the lines due to spin correspond?

Question: How did each electron acquire its particular direction of spin? Can +½ spin electrons be separated from –½ spin electrons?

Comment: Science has not explained electron structure, and tries to distract us by arguing there is no need to consider it. From the discussion above, it is impossible to believe there is not some kind of structure. Recall the Einstein speculation (See Ch. 10. Fitzgerald-Lorentz Contraction. § H.).

Let's go back to the observations. The total deflection in the actual trace was only 0.22 millimeters (!).

Comment: Ideally, the trace would have been two curved lines. Since the two bands in the trace were of undisclosed width, it is remarkable that a space between them was detected.

The diagram on the printed page in **CVEP** shows a maximum deflection of about 4 mm, with a separation of 2 mm between the inner edges of the two bands. If this is to scale, then the separation observed would have been only about 0.11 mm. And this is with a powerful magnetic field.

4. Summary view

• The concept of spin is bizarre: there is a point that has no dimensions and yet has properties due to imputed rotation about an invisible axis through the point.

• If the Bohr model of the atom is inadequate, and must be replaced by a wave model, then all the derivations and interpretations based on the Bohr model, including Rydberg, must be discarded.

B. Particles

All except mediating particles are grouped by mass into light (leptons) and heavy (hadrons) particles. There does not seem to be a continuum of particle masses, otherwise Science would have a problem with defining the boundaries between these groups.

1. Leptons

Table 11.4 shows the leptons. The neutrinos have no charge; the other particles listed are all −1. There are also 6 antiparticles, with presumably the same properties except for opposite charge. All have spin ±½. This seems to mean that a lepton can show up with a +½ spin or a −½ spin. There are equal numbers of the two species. I have not found a statement that a lepton with a positive spin can or cannot change into one with a negative spin. Some writers use the term "life" instead of "half-life". Since these apply to decay reactions, "half-life" seems more appropriate.

The muon and the antimuon were found in cosmic rays that are composed of high speed particles from far distant stars. Yet the muon life is only two millionths of a second. The muon cannot decay into 212 electrons to conserve mass because charge cannot be created. The muon (mass 106 MeV) decays into an electron (mass 0.5 MeV), a muon neutrino, and an electron antineutrino. The remaining 105.5 MeV adds to the kinetic energy of the products. The Law of Charge Conservation does <u>not</u> explain:

• why the muon breaks down;
• how it managed to last till it got to Earth;
• whether some process is creating muons near us.

Table 11.4 Leptons

Particle	Mass, MeV		Half-life
	EST	**TEU**	
Electron	0.511	0.5	Stable
Electron neutrino	< 60E–06	< 13E–06	Stable
Muon	106	106	2.2 µs
Muon neutrino	< 0.6	0.25	?
Tauon	1800	1800	0.3 ps
Tauon neutrino	< 150	< 35	?

Comment: Conservation of muon number means there is some characteristic of a muon (muon-ness) that must be conserved even in a massless neutrino, so it logically is not connected with mass. Similarly for the tauon, which decays in 0.3 trillionths of a second. Science doesn't know what muon-ness and tauon-ness are.

2. Hadrons

The particles included in hadrons are heavy (baryons), intermediate (mesons), and ultraheavy (hyperons). The only stable hadron is the proton. The free neutron has a half-life of 10.6 minutes. All the rest have vanishingly short lives, measured in no more than millionths of a second down to trillionths of trillionths of a second. Hadrons are subject to strong and weak interactions. All have antiparticles except the neutral sigma particle, which does double duty.

It would seem the concept of Conservation of Baryon Number was invented to explain why some reactions occurred or did not occur. Explain? No. It was invented to save the theories, like the epicycles of Ptolemy.

3. Protons/neutrons

A neutron decays thus:

$$n^0 \rightarrow p^+ + e^- + \text{antineutrino}^0 + \text{photon}^0$$

The mass of the antineutrino is 0; the photon mass is unsettled. The mass balance in units of E–24 grams is:

Neutron mass		1.6749543
Proton mass	1.6726485	
Electron mass	0.0009110	
Total		1.6735595
Mass loss		0.0013948

The mass loss is equivalent to 0.782 MeV. This confirms that the calculation is based on rest masses, and assumes that the neutron, proton and electron have negligible kinetic energies. If we can measure the photon energy to within 10%, the mass loss is then uncertain by 1.395E–28 g, which can be taken as the maximum possible mass of the antineutrino, and presumably the neutrino. It is 1/6th the electron mass.

> **Question:** Why does the neutron decay? Because it is unstable. That explains nothing; it only labels the observation. The further mystery is that it is unstable only when isolated; in association with protons and neutrons in a nucleus it is stable. Except for radioactive decay. Is the proton part of the neutron trying to reject the electron or is its attractive power too small to hold the electron without help from at least one other proton?

We reverse this to make a neutron: we provide a high energy proton and a high energy electron plus an anti-neutrino. Or perhaps a proton and an electron yield a neutron plus a neutrino. Either extreme pressure or extreme energy, or both, can force an electron to combine with a proton. The need for a neutrino appears to be a reason why an electron ordinarily doesn't get closer to the proton than orbit 1.

> **Comment:** This is strange: the attraction between the proton and the electron is powerful. It is as if the proton were saying to the electron, "Come closer, so I can reject you." It is in sharp contrast to electron/positron annihilation.

Actually, both are responsible for the attractive force, and so both are saying this to the other, but are thwarted by some other force which makes their union difficult.

But if we add another proton, the two protons and the electron can exist as a proton and a charge-neutralized neutron. This is the deuteron.

> **Question:** When a neutron forms, what happens to the charges? They are apparently available to be reconstituted when the neutron decays.

The electron may be 1/10 the size of the proton or 2000 times larger. Does one envelop the other? Is the stuff of one squeezed onto the other as a layer? Do they merely remain as tangent spheres? Or are there neutron waves?

> **Comment:** The antiproton has the same mass as a proton, but a negative charge equal to that of an electron. Apparently the negative charge can be spread out over a larger volume than that of the electron. And the positive charge of a proton can be condensed on a positron.

4. Antineutron

AJASC notes that in 1956 in a close encounter, without collision, of a proton and an antiproton both lost their charges. The "explanation": the proton became a neutron and the antiproton became an antineutron.

> **Comment:** How? The proton mass is less than that of the neutron, so the mass gain must have come from kinetic energy. **AJASC** says the difference between a neutron and antineutron is that while their spins are in the same direction, their magnetic fields are in opposite directions. This strongly suggests the particles have structure.

C. Fundamental forces

Table 11.5 shows fundamental forces and related particles. **EST**, Elementary Particles, says a particle in relativistic quantum mechanics is an isolated system with a mass and a spin. Since gluons and quarks are never isolated, they are not particles. **EST**, Mesons, says that gluons are neutral particles. They are included here because they seem to represent a force.

Table 11.5 Fundamental Forces and Exchange Particles

Force	Mediating Particle	Mass, GeV	Charge
Gravity	Graviton	0	0
Electromagnetism	Photon	0	0
Strong nuclear force	Pions	0.140	±1
		0.135	0
Weak nuclear force	W	82	±1
	Z	93	0
Intrabaryon force	Gluons	0	0

1. Gravity

Gravity is a real challenge: its effect is felt over vast distances with nothing in between. The transmission is supposedly by rapid exchange of gravitons, which have no mass, no charge, are different from neutrinos, but have never been found. We ask:

• Instantaneously? Across millions of light-years? Are there distant bodies that have not yet felt the gravitational attraction of our galaxy?

• "Exchange" means our Sun receives gravitons from our Earth. What drives them?

• How many gravitons are there per unit mass?

• When matter appears, is its gravitational force felt immediately throughout all space? Or does its influence gradually spread, but once spread remains undiminished?

• Our Sun converts enormous amounts of mass to energy. Before conversion, this mass emitted gravitons that are speeding on to other bodies. Right after conversion, do these suddenly disappear? And what of the gravitons it was receiving from other bodies?

• Do gravitons cause the reported bending of starlight passing by our Sun? What happens after the interaction? Science may talk about gravitons because of an intellectual craving for symmetry in interaction mechanisms.

2. Strong nuclear force

The strong nuclear force was imagined to "explain" what holds nuclear protons together against the strong force of repulsion between like charges. Also to explain how electrically neutral neutrons are held either to each other or to protons.

Comment: Let us look at how large this force is. Imagine two protons as spheres in direct contact. Treating their charges as point charges concentrated at the centers of each sphere, Coulomb's Law gives us (Atomic Models, § C.2.):

$$F = A / (2.8E–15 \text{ m})^2,$$

or 29.4 newtons. Now the projected area of each sphere is $(\pi/4)(2.8E-15$ m$)^2$, or 6.16E–30 m^2. So the force per unit area 4.78E30 newtons/m^2, or 6.93E26 psi. Truly enormous. **CVEP** says the strong force is 100 times stronger. How is this known?

Next, Science asserts that the reason the nucleus stays together is that the protons and neutrons are constantly exchanging something. What? The strong force is said to decrease inversely as the seventh power of the distance. Its effective range is about 1E–13 cm, which is about 70% of the radius of a proton. So a proton or a neutron can interact only with adjacent particles.

But there should be corresponding enormous attractive forces from the "electrons" in the neutrons, whether or not they exist as electrons.

It is pions that are being exchanged. Powell in 1947 in studies of cosmic rays reported tracks whose curvature correponded to particles having a mass 273 times that of an electron, making them mesons. They were called pions, and show +, –, or 0 charges. The neutral pion has no antiparticle. The pions were said to carry the strong force. What does the neutral one do?

The descriptions don't state how many pions per proton-neutron or proton-proton pair there are. What happens in atoms with unequal numbers of neutrons and protons?

The pion half-lives depend on whom you read. Charged pion lives are about 30 billionths of a second long; the neutral pion, 15 quadrillionths of a second. The charged pions decay to muons and neutrinos. Muons decay to electrons and neutrinos. The neutral pion decays into gamma radiation (two photons).

3.   Weak nuclear force

This was imagined to help "explain" neutron decay. **CVEP** says its strength is ten quadrillionths of the strong force; **AJASC** says less than ten trillionths. Its range is one thousandth the diameter of the nucleus. (For hydrogen this is 28E–15 cm). It does not hold anything together (!): it "mediates" particle conversions, e.g., protons combining to helium nuclei. It is carried by W and Z "exchange" particles, for which, somehow, are predicted masses of about 100 GeV.

Comment: These particles don't prevent the decay of either a free neutron or neutrons in radioisotopes. Decay may involve a repulsive force. In stable nuclei, they may be a glue for neutrons, because the

197

range of the weak force is confined to the outer portions of a neutron or proton.

Comment: The energy equivalent of the rest mass of the proton is 0.938 GeV. So how can particles with 100 times the mass of a proton be tolerated within the nucleus? None of the calculations of nuclear reactions I have seen show W or Z particles.

We are told a handful of events in particle accelerators out of over a 100,000 were identified that seemed to confirm the W particles. We visit the Wonderland of Quarks next.

D. Quarks

   1.   Characteristics

Science postulates other particles called quarks. Their masses range from 1/200th to 200 times the mass of a proton. They are inferred from particle tracks in collision experiments that are interpreted as evidence of interactions of quarks. It is speculated that we will never see a free quark.

Comment: This is clever. Religion is attacked for its faith that God visited Earth and ridiculed for believing His promise to come again. Science avoids this difficulty by constructing its faith so as to eliminate the hope of ever seeing a quark. There is a consequence: it is an indirect attack on the concept of hope.

A complex system has been developed around the notion that there are 6 quarks, listed in Table 11.6, and their 6 antiquarks. We are not going very deeply into quarks. The trail is very difficult to follow because the publications I consulted gave partial descriptions, unintelligible to nonphysicists.

The three pairs shown are called "flavors". Scientific whimsy did not settle on names like vanilla, chocolate and strawberry, or salt, sweet and sour. The flavors differ in mass, so why not small, medium and large? The names of the first and third pairs suggest opposites, but their charges don't reflect this.

Question: Since quarks have never been seen, how was it established that they have any spin at all? Perhaps it was agreed they were fermions and thus to be consistent they had to have a spin of ½.

The mass of the top quark may be greater than 205 proton masses. Recently,

Table 11.6 Quarks

Name	Charge	GeV[a]
Up	+2/3	0.003 - 0.005
Down	−1/3	0.004 - 0.010
Charmed	+2/3	0.77 - 1.5
Strange	−1/3	0.077 - 0.200
Top	+2/3?	24? - 192?
Bottom	−1/3	2.5 - 5

[a]1 GeV = 1.60219E–03 erg $\approx$ 1.7827E–24 g
     = 1.06584 proton masses
The sign $\approx$ means "is equivalent to".

it was announced that the top quark has finally been found. We could say, "Uh, huh." But reflect on this.

• If it really is a building block, there should be more of them around. What is its function? Why was it so elusive? Because of its extremely short life and multiple decay modes, we are told.

• Remember, we are searching debris. Was this particle singled out because it fits some model? Where do the some two hundred other particles found fit?

2. Quarks in hadrons

The hadrons are made of quarks and antiquarks (Table 11.7). The attractive force between quarks <u>increases</u> with distance!

Comment: Now <u>this</u> is a bold departure from familiar inverse square laws for forces and all our other experience which shows that the interaction between two bodies continually grows less as they separate. Is it linear with distance? To what distance does it hold? How is this property demonstrated since we've never seen a quark or a gluon? It suggests a rubber band or a spring.

The omega minus particle has three identical quarks, which violates the Pauli exclusion principle. Science invented another property, color, in three

Table 11.7
Quark Makeup of Some Hadrons

Baryons: All are made of 3 quarks
    Proton   uud
    Neutron   udd
    Xi minus dss

Hyperons: All are made of 3 quarks
    Lambda 0   uds
    Omega minus sss

Mesons: All are made of a quark + an antiquark

Particle		Strangeness Number
Kaon		
+1	us*	+1
−1	u*s	−1
0	?	
Pion		
+1	ud*	0
−1	u*d	0
0	?	

- u = up; d = down; s = strange
- Letters with asterisks are antiquarks

---

varieties, to remove the violation: red, blue and green, giving us 36 quarks. "Color" is admittedly meaningless.

The justification offered in **AJASC** for color is that one of each color produces white, as in photography. No hadron is colored so the three quarks in each of them must be of different colors. No meson shows color either, because all are made of a quark and an anti-quark, and apparently this also leads to "white".

Leptons are not made of quarks. Does <u>any</u> particle show color?

Comment: Science copies Religion in having Trinities: 3 flavors, 3 colors, 3 kinds of leptons and 3 kinds of lepton neutrinos. Creationists might say God is nudging the rebellious to think in terms of Trinities.

3. Gluons

The neutron and the neutral lambda particles have two negatives and a positive. The xi minus and the omega minus have three negatives. So the "theory" required a further assumption: there was a glue holding such particles together. Remember Ockham?

So, all quarks are held together by gluons. There is one gluon for each generator of the symmetry group, whatever that means, for a total of 8. Gluons have color and can change the color of a quark (but do they in every reaction?).

Comment: AJASC writes as if gluons are particles when inside a hadron. Or are they waves? Or just some kind of schmier? So far Science has assumed only two kinds of things in space: particles of matter and waves (wavicles? partiwaves?). Notions of symmetry and stability would argue for at least a third component. Is space itself the third?

Comment: The electron (and perhaps the gluon) has the character of an immaterial spirit when in an atom. Outside the atom, it is incarnated as a particle. Science again seems to borrow frequently from the concepts of Religion.

Now for the next assumption: gluons don't hold quarks together on their own accord. They operate through intermediaries, the pions! But the pions are also made up of quarks, which are held together by gluons operating through pions. This is fascinating, so let's go a bit further down this path.

The proton has two possible arrangements. For simplicity, let's label each quark by its color, listing the two up quarks first: R-G-B and R-B-G. Both have counterparts with antiquarks in place of quarks. At every bond are gluons. Must all the gluons in a hadron be different? We have only 8, so let's assume the gluons on either side of a quark may be the same.

a) R-1-G-1-B   b) R-2-B-2-G

B may bind back to R via a type 2 gluon in a) and G may bind back to R via a type 1 gluon in b). But this is not the end. Each gluon is madly exchanging pions with both quarks it is binding. Which pion is it? It might help to use a

201

negatively charged pion to bind two positively charged up quarks, and a neutral pion to bind a positive and a negative quark.

Let's say each gluon can utilize only one type of pion. Next, each pion is itself made of quarks. From Table 11.7, the positively charged pion is made of an up quark and a down antiquark. So what color are these? Let's say they have to be different from the one they interact with. Using an asterisk to indicate an anti-quark, and adding a final hyphen to bind back to the initial R, the R-G-B proton becomes:

-R-(g-r*)-1-(g-r*)-G-(b-r*)-1-(b-r*)-B-(g-b*)-2-(g-b*)-

This still doesn't meet the rules, because we have quark (e.g., R) against quark (e.g., the g in g-r*). Don't we need a gluon there?

-R-3-(g-r*)-1-(g-r*)-3-G-3-(b-r*)-1-(b-r*)-3-B-3-(g-b*)-2-(g-b*)-3-

i.e., quark-gluon-(pion)-gluon-(pion)-gluon-quark-etc. We see this process has no end, for the two quarks in each of the pions are held together by other gluons.

Comment: Even if the process can be defined to have an end, how many "particles" can one stuff into a proton, which does have a definite unchanging size?

Comment: When we turn to mesons, of which pions are one species, each has only two quarks. The gluons are very adaptable in being able to bind either three or two quarks.

4. Charge

The electron has been considered as the smallest particle with a unit negative charge. Now we are told that there are particles that can be treated as having one-third of this unit charge. So the indivisible is divisible, except when Science says it's not.

The unit positive charge is also divided into two charges of +2/3 and one of −1/3.

Comment: The logic earlier was that all negative charges found were integral multiples of that on an electron so the latter was set at −1. Why isn't the charge on the down quark now set at −1? This would make the charge on the electron −3, and that on the proton +3. It must be less upsetting to use fractional charges for quarks since they don't appear in normal space and the nasty fractional charges can be buried in subspace. And why isn't the electron made of two –2/3 charges and one +1/3?

5. Quark forces

Dirac estimated the diameter of a quark as 1E–15 x that of an atomic nucleus, or 280E–30 cm. As a sphere, its volume is 1.149E–83 cm^3. The smallest quark has a mass of 0.005 GeV (8.91E–27 g), so its density is 7.76E56 g/cm^3. This is 15 quintillion sextillion times as much as an average neutron star, which arouses interest in calculating the force between two tangent up quarks.

Calculation: The calculation is similar to that in § C.2. Here, the charge is 2/3 and the center-center distance of the quarks is 2.80E–28 cm.

$$F = (2/3)^2 \ A/R^2$$
$$= 1.308E36 \text{ dynes.}$$

This is a very large force, but it needs to be seen as operating on the very small projected cross-sectional area of the quarks.

$$F/A = F/[(\pi/4)(2.80E\text{-}28 \text{ cm})^2]$$
$$= 2.12E91 \text{ dynes/cm}^2,$$

or 2.1E85 x the pressure of the Earth's atmosphere. For the gluons to overcome the repulsive force of two positive quarks seems not credible.

Comment: Some say we cannot hope to investigate the structure of a quark. However, merely postulating the quark has brought us full circle back to Anaxagoras and Euclid who held that there was no limit to subdividing particles.

E. Particle lives

Floating around in our Universe are only three stable particles: the proton, the electron, and the electron neutrino, if it exists. Some would add the photon. All the rest have extremely short lives.

As research continued, Science "discovered" many more particles. The way this happened was to bombard particles with high energy neutrons, protons, or electrons and examine the debris. They created the new "particles", rather than freeing them from other particles.

Comment: Now suppose we erected 100 buildings using standard bricks (2" x 4" x 8.25") made by the same manufacturer, in the same kiln, of the same materials, by the same process and using the same workmen under the same working conditions and cement from a single source. Let the buildings cure for a year. Now let us demolish the buildings with a wrecking ball. Would we not expect to find among the various fragments some similar fragments? And yet all are ultimately made of the same unit brick. What significance do the larger composites have? None. For particles, one might argue that whether intermediates can even form might give clues as to the nature of the binding forces and structure.

Comment: Even if a particle of a particular mass can be repeatedly demonstrated in collision experiments, such particles do not automatically qualify as fundamental building blocks.

However, to explain our Universe we need something more detailed than the simple proton-electron-neutron model.

Bizarre Concepts

Physics offers even more bizarre conceptions.

- Superstring theory, in which particles are replaced by one-dimensional strings created out of vacuum that have quantized oscillations.

  Comment: The transverse dimensions of a line are zero. A Universe cannot be built by adding zeros. So "one-dimensional" seems to mean only that the transverse dimensions are microscopically small compared to their length.

- 26-dimensional Universes with curled up vanishingly small sizes for most of the dimensions.

  Comment: What are the variables being measured in these various dimensions? Can our normal three spatial dimensions curl up into microscopically small balls? This is nonsense, because there is no physical line in any direction to curl up. Well, it is "explained" that these three are exempt from curling up, meaning only that this exception is needed for the concept to work. The objection holds for the other 22, assuming time doesn't curl up too.

- Magnetic monopoles: isolated north and south magnetic poles

Comment: This violates Maxwell's theories. One monopole is said to be 10 quadrillion times as massive as a proton, which is still small: about 17 nanograms. They have never been detected, and by some theories should be rare.

- Axions: particles whose mass is 1 billionth that of an electron. They are different from quarks.

- Virtual particles: particles that appear spontaneously out of a vacuum and disappear back on a time scale too short for Science to make measurements on them.

  Comment: Cleverly preventing a check of their existence by the use of the main tool of Science.

- Vacuum energy: even a vacuum has energy.

One can admire the imaginativeness of these mathematical constructs, but eventually they have to be related to the real world. They also need to be translated from the esoteric language of physics. One may suppose that with several years of study one can begin to follow some of the equations, but for most the material will remain impenetrable, incomprehensible, and probably irrelevant.

Irrelevant because we are in need of something much more important than an understanding of the structure of the Universe and its mechanical workings. And that something is being ignored by Science. Some even wish to persuade us that some questions need not be investigated. **GNP** says things like:

- "It is often necessary to resort to abstract mathematical formulations to make sense of the world. Ordinary experience can be an unreliable guide."

  Comment: Both statements may be true in some cases, but let us be aware of the underlying danger in the exhortation to rely more heavily on the abstractions of mathematics rather than on those of our experience or common sense.

- "Human intuition is an inadequate guide; we must use mathematics."

> **Comment:** Human intuition has often given us answers not deducible by logic: invisible atoms, gravity, bacteria. Also, how many discoveries were the result of accident? Logic is not a substitute for intuition or the capacity to respond to the unexpected. It is rather a tool for intuition and curiosity.

• "Failure of the human imagination to grasp certain crucial features of reality (?) warns us that we cannot base great religious truths on simple-minded ideas of space, time and matter gleaned from daily experience."

> **Comment:** Now we are to believe that mathematical abstractions are not merely representations but <u>are</u> reality. Our perceptual capacities are denigrated as simple-minded, and the lessons gleaned from thousands of years of human experience are discounted and discarded. Why single out religious truth? The statement applies to all proposed "truths", including "scientific truths".

• "The arrival of a particle at a particular point is inherently unpredictable and thus has no cause."

> **Comment:** An appalling non sequitur. Because we cannot predict it does not mean it has no cause. We cannot predict earthquakes but they have causes. We know prices on the stock exchanges fluctuate. We cannot predict them, even though we know at least some of the causes.

Arguments such as these should make us wary. We are being asked to submit to new masters who "know" more than we do and to let our critical faculties atrophy.

Overview

What have we found in the world of atoms?

• Science is frustrated in trying to peer into sizes that are smaller than its capability of probing.

• A serious obstacle is the Heisenberg Uncertainty Principle, which <u>guarantees</u> that we will never even partially understand the microcosm because Science confesses it cannot be measured precisely.

• Quantum theory does not apply to events whose scale is larger than the atomic.

- Goedel's Proof (mathematics can never construct a system that can be proven to be internally consistent or inclusive of all truth) is being ignored.

- Ockham's Law, arguing for a minimum of assumptions, is being ignored.

- Science has not offered any clue as to what will be done with any answers it finds as to the origin and structure of our Universe.

So I do not feel it is necessary to pursue this path further because in all the postulates so far I have found only total avoidance of questions such as:

- Where did it all come from?
- Where did it start?
- How did it start?
- Why did it start?
- What was going on before it started?

In fact, I am moved to irreverence. On the one hand, we are told that space is curved.

- One way gives us a closed Universe, like an egg.

- Another gives us an open saddle-shaped Universe, whatever that means.

- Still another gives us two open cup-shaped Universes separated by an unknown distance.

On the other hand, we are told that space is made up of countless strings, one-dimensional things that do not intersect. My irreverence comes in thinking we are being told we live in a bowl of spaghetti, with possibly an invisible sauce of dark matter.

So let us turn our gaze away from the submicroscopic world of atoms and go back to looking at the stars.

Chapter 12. Nature of Our Solar System (Contd.)

We first return to examining our Solar System. Our objective is still to determine whether what Science teaches us about it is definitive and credible.

Mass of the Earth

Science cannot put the Earth on some cosmic scale, but Newton's Law of Gravitation tells us its mass.

$$F/M_2 = GM_1/(R_1 + R_2)^2 \approx GM_1/R_1{}^2 = g$$

M_1 is the Earth's mass, M_2 is any other spherical mass on the surface of the Earth, and the R's are the radii of the two masses. Their sum is the distance between their centers of mass. We can make M_2 so small that R_2 is negligible compared to R_1. F/M_2 is the acceleration, g, given to mass M_2 by force of gravity F.

We can measure both g and G in the laboratory, so we can solve for M_1. The Earth is not a perfect sphere, but the current value for the equatorial radius is 6.3782E08 cm (**TEU**, Appendix A). The mass is 5.979E27 g; the same reference gives 5.977E27 g.

Density of the Earth

As a perfect sphere, the Earth's volume is 1.0869E27 cm^3. Dividing this into the mass gives an average density of the Earth as 5.50 g/cm^3 vs 1 g/cm^3 for water. Most rocks are no more than 3 g/cm^3. Something in the Earth must be much denser than 5.50.

Structure of the Earth

Earthquakes generate shear (S) waves and compressional (P) waves. Liquids and solids can transmit P waves but only material that has some rigidity can transmit S waves. Seismic waves seem to be the <u>only</u> means for probing the deep internals of the Earth.

A. Regions

From study of these waves, it is postulated that the Earth consists of three concentric spherical shells around a solid core. Some think the core is liquid.

• The core is about 1280 kilometers (km) in radius, and is a mixture of metallic iron and nickel.

Comment: Science has no core samples, so it takes a hint from meteorites. These are stones of various sizes containing oxides of various elements. About 10% are made of elemental iron (density 7.9 g/cm^3) with some nickel (8.9 g/cm^3). Perhaps they came from a planet that broke up. But perhaps they were expelled from the Sun and are evidence of nonuniform reactions in it. Their greater densities let them sink through the molten solids comprising the rest of the Earth. But how did they escape oxidation? Or did they oxidize and then on reaching the hot center dissociate into elements again or get reduced by carbon? This could mean our oxygen atmosphere was not captured later from space but was outgassed from within the Earth.

Science cannot <u>positively</u> identify the source of meteorites. Every year the Earth plows through clouds of meteoroids at various times in its orbit. Any meteorite is thus a chance event in space.

• The core is surrounded by a liquid region about 2200 km thick. Some call this the outer core.

• Above this is the mantle, about 2870 km thick, which is largely solid but apparently contains liquid regions.

• Finally, there is a crust which is about 35 km thick under the continents and about 10 km thick under the oceans.

B.   Effect of pressure

The pressure on the central core due to the "weight" of 5105 km of material above it is approximately:

$$[(5105 \text{ km})(3281 \text{ ft/km})] \, [(62.4 \text{ lbs/ft}^3)(3)] \, / \, 144,$$

or 21.8E06 lbs/in^2. This is 1,500,000 times normal atmospheric pressure at sea level. The factor of 3 is an assumed average density relative to water.

From the laws of heat conduction, the center of the Earth <u>cannot</u> be cooler than any other point. **TDU** says the temperature is in the range 6000 – 6400 K. This is about the surface temperature of our Sun. The crust and mantle are solid since they are neither hot enough nor compressed enough to be liquids. The outer core may have a different composition than the central core, or some unknown factor may permit the central core to appear solid. Thus, we stand on a solidified scum of rocks floating on a denser liquid.

Charles H. Peterson

<u>Age of the Earth</u>

A. Human records

How do we know how old the Earth is? At least as long as humans have been around. The Christian Bible has yielded an estimate that God created the Earth in 4004 B.C. Written records go back six thousand years. Artifacts and cave drawings are dated back 10,000 - 50,000 years to the Upper Pleistocene. Neanderthals (Mousterians), now classed as human, date to 60,000 - 150,000 years ago in the Middle Pleistocene. I don't know how the dating was done.

There are also mistakes and frauds. Dr. Eugene Dubois in 1891 discovered a part of a skull, a thighbone and three teeth that became known as Pithecanthropus erectus, dated to 750,000 years ago, or Lower Pleistocene.

Dubois himself later decided the fossils came from a gibbon. Piltdown Man was a fraud. Nebraska Man was contrived from one tooth, that eventually was identified as from an extinct pig (**TCOE**).

B. Natural records

1. Tree rings

Trees seem to produce two rings every year. They are not uniform in width and may be wide or narrow, perhaps reflecting variations in rainfall. The patterns in trees of one period can be matched with those in trees from excavations to extend the calendar back in time. The longest chronology thus far goes back about 8000 years (**EST**, Dendrochronology).

2. Lithological cycle

The geologic features of the earth are due to erosion, stratification, and crystallization of complex rock minerals, all of which appear to require very long periods of time to occur. How time scales were established for these processes is not clear.

Rocks are classified as igneous, sedimentary, and metamorphic. It appears that a cyclic change from one type to another is possible.

• Magma (molten material) cools to igneous rocks having large amounts of silicon dioxide (silica).

• Gradually the igneous rocks are changed chemically, structurally or both to metamorphic rocks under the influence of water, heat, pressure, and other minerals, such as silicates and carbonates.

• All rocks can be degraded and broken down to particles by expansion and contraction cycles due to temperature variations; freeze-thaw cycles involving water and ice; wind and water erosion; and leaching by water. The particles are moved by air, water, or ice to low lying areas such as stream beds where they form sedimentary rocks by compaction from the weight of overlying material and by chemical cementation. These processes are generally very slow. Sediments in the Mississippi delta are at least seven miles thick. About two million tons of sediment are deposited per day (**CAE**). The time for deposition is more than 100,000 years.

• Through tectonic movements, both metamorphic and sedimentary rocks can be subducted into underlying magmas and eventually recycled as igneous rocks.

C. Radioactive decay

    1. Radioisotope dating

The radioactive decay of elements shown in Table 12.1 is used for dating specimens. The half-life is the time required for spontaneous decay of one-half the number of original atoms. Exhibit 12.1 shows a simple relationship that appears to adequately represent this decay. It is a statistical law: though we cannot tell which atoms will decay at any one instant, on average a certain fraction will.

The application to dating is not as straightforward as might be hoped. The basic requirement is that the specimen has been an isolated system: nothing entered or left after the specimen formed. Consistency in the observations is the criterion for their interpretation.

    2. Isotopes

Our Earth may once have contained 92 different substances called elements, distinguishable by their size, mass, and chemical behavior. Each element has a fixed number of nuclear protons, but may have a range in the number of neutrons. Each combination is an isotope. E.g., the range of the uranium series is U-226 to U-240. U-238, the principal member, has 92 protons and 146 neutrons.

Table 12.1 Some Radioisotopes Useful for Dating

Element	Half-life, years	Time span, years
Carbon-14	5.73 E03	Back 40 E03
Potassium-40	1.3 E09	Back 2.8 E09
Uranium-238	4.5 E09	Back 3.8 E09
Rubidium-87	48 E09	Back 3.8 E09

Most nuclear combinations are unstable and are called radioisotopes. They spontaneously emit energy, mass, or both in changing to the stable isotopes we find in nature. These will be elements completely different from the parent nuclides, e.g. U-238 decays through 14 steps to nonradioactive lead-206. Seven are found only because they occur in decay chains of long-lived elements. Two others have only radioisotopes with relatively short half-lives: technetium (43 protons) and promethium (61 protons). They can be man-made.

3.   Half-lives

The range of half-lives is from billions of years to picoseconds (trillionths of seconds).

> **Comment:** Apparently the scientific community has decreed that if something lasts for picoseconds it shall be granted the status of existence. This is reasonable if to get from nuclide A to nuclide C a substance <u>must</u> pass through short-lived nuclide B. The existence of nuclide B helps understand (not explain) the decay process. But if A, the proposed first member of the series, has a vanishingly short half-life, what does it mean to say that A exists?

At the other end of the time scale are nuclides whose half-lives are so long that they could have been in existence before the Earth came into being. Examples are tellurium-128 (1.5E24 years), tellurium-130 (2E21 years, and selenium-82 (1.4E20 years). Science says these are products of supernova explosions.

> **Question:** Religion might ask why did God make them before Earth? Excepting selenium, none are of any use to human existence. Of all Se, 91% is nonradioactive (Se-80, 78, 76, 77, and 74). **Answer:** Why does a carpenter make sawdust and wood chips?

There is no proof that the "stable" elements do not decay, but they are treated as if they do not.

4.   Indications of age from technetium-99

We have no reason to suppose that equal amounts of all isotopes were made originally, but it is not unreasonable to suppose that <u>some</u> amount of each was made.

Rubidium-99 (Rb-99) is a fission product of uranium, decaying rapidly in six steps to technetium-99 (Tc-99), and then finally to stable ruthenium-99 (Ru-99). Since the half-life of Tc-99 is 214,000 years, the decay chain in effect begins with Tc-99. No Tc-99 is found in nature. Explanations include:

- Neither Tc-99 nor its precursors ever existed.

- There never was very much Rb-99 so all members of the decay chain have had time to decay completely.

- Ru-99 could have come from cosmic ray (alpha particles) bombardment of stable Ru-96.

The upper limit on the original amount of rubidium-99 may be estimated as 1.3 ppb, based on decaying to an amount of Tc-99 undetectible today. The time is about 13 million years. For Tc-98 (half-life 4.2 million years), the time increases to 220 million years.

5.   Uranium decay

Natural decay chains account for 98.5% of natural lead.

- Uranium-238 decays via radium-226 to lead-206.
- Uranium-235 decays via radium-223 to lead-207.
- Thorium-232 decays via radium-228 to lead-208.

We can assume that the Creator initially made:

1) only the parent nuclides which then decayed to the elements we find (radium and lead); or

2) parents and intermediates in just the right amounts to produce the ratios we find; or

3) parents and intermediates and is continuing to make all of them.

Assumption 3 means we can never make any sense of our Universe. However, we have observed that our mined out gold or copper mine is <u>never</u> spontaneously replenished. We have come to expect that this is true for all minerals. Assumption 3 is unlikely to be true.

In Assumption 2, most of the isotopes of radium have short half-lives measured in days down to microseconds. Consider Ra-228, which has a half-life of only 6.7 years. In the roughly 900 half-lives since Creation, it would have decayed by a factor of 2^{900}. This works out to an original mass 10^{243} times the present mass of the Earth. The radium-228 we find must be the result of decay of Th-232, and we forget about Assumption 2.

> **Comment:** There is no mention of radioactivity in the Bible. This means that even if the Bible is infallible, it is not comprehensive. Of course, it is focussed on relationships, not technology.

Assumption 1 permits estimating the age of the Earth. Creationists insist the Earth is certainly no more than 10,000 years; Science says 4.5 billion. Now radium is found <u>only</u> in uranium/thorium minerals and lead is <u>always</u> found in all such minerals. Lead is, of course, found in other minerals. Some may be primordial lead, but some may have leached from radioactive ores and been converted to other chemical forms.

Table 12.2 shows expected (calculated) ratios of decay products to parent nuclides for two different time periods. Column 2 shows that in 10,000 years, practically none of the original 1 g of U-238 has decayed; Column 4 shows half has decayed in 4.5 billion years. To show clearly the increase in the relative amounts of the daughters, we divide each of these ratios by the amount of U-238 remaining and multiply by 1E09 to get parts per billion of U-238 remaining. The ratios found in nature are (**NCE**):

Th-230   1.76E+04   Ra-226   3.53E+02

These agree quite well with those calculated for 4.5 billion years of decay.

> **Comment:** Some Creationists object that the decay constant is not constant. Science has conceded that in a few cases found thus far there is a small range in the values of the constant. As noted in Exhibit 12.1, even the severe conditions in a nuclear reactor do not affect the constants. And even a 10-fold increase in k for U-238 still means a decay time of 450 million years. Of course, if k were smaller, the decay time would be longer.

Table 12.2 U-238 Decay for Two Time Periods

Nuclide	10,000 y		4.5E+09 y	
	Ratio	ppb	Ratio	ppb
U-238	1.00E+00	1.00E+09	4.98E–01	1.00E+09
Th-230	2.11E–08	2.11E+01	8.40E–06	1.69E+04
Ra-226	2.90E–10	2.90E–01	1.78E–07	3.58E+02

6. Rubidium decay

We need to look a little into the decay of rubidium-87 (Rb-87) because it is used to establish the age of meteorites that are said to be as old or older than our Earth. This is so important that it merits inclusion here in the main text rather than in another Exhibit.

The Rb-87 nucleus ejects an electron to become stable strontium-87 (Sr-87). Its half-life of 48 billion years yields a decay constant of 1.444E–11/yr. In 4.5 billion years, only 6.3% has decayed.

Data: EST says the Earth's crust contains 90 ppm of rubidium and 375 ppm of total strontium isotopes. **SSE** says 32 and 260. The averages are 61±29 and 318±58 ppm. From isotope tables we find that Rb-87 is 27.7% of the total and Sr-87 is 7.0%. Thus we now have 17±8 ppm of Rb-87 and 22±4 ppm of Sr-87, so originally there was 39±12 ppm of Rb-87.

Comment: This is inconsistent: more than half the Rb-87 has decayed, not 6.3%. Either we have lost Rb-87, gained Sr-97, or started with more Sr-87.

With no loss of Rb-87, the initial amount was 17/0.937, or 18.1 ppm, and there should now be only 1.1 ppm of Sr-87. The third possibility seems most likely. What was its source?

U-235 and U-238 yield small amounts of Rb-87 by fission in power reactors. This happens in nature to less than 1% of these radioisotopes in half-lives longer than a quadrillion years. Thus, the Rb-87 must have been essentially all primordial.

The technology of dating is more complicated. The radioactive decay law can be manipulated into:

$$[Sr\text{-}87]_t = [Sr\text{-}87]_0 + [Rb\text{-}87]_t\,(e^{\lambda t} - 1),$$

where the terms in [ ] mean the number of atoms of the isotope shown and the subscript t means the number is at any time t. Instead of counting atoms, a mass spectrometer is used to measure the ratio of each isotope to a reference isotope, as Sr-86 in this case. Each term in the equation is thus divided by $[Sr\text{-}86]_t$. This is 9.8% of the total strontium, or 31 ppm. Any precursors have long since decayed by electron emission, so this amount is essentially $[Sr\text{-}86]_0$.

Procedure: Measure the ratio of the number of atoms of the two isotopes at time t in several samples of rock that are judged to have the same geological age and plot a graph of one ratio against the other. A straight line called an isochron should result. Its slope is the term in parentheses, from which the age t is calculated.

Since the atoms might migrate, whole rock analyses are used rather than mineral analyses, because the whole rock is more apt to have remained a closed system and be representative of the original chemical composition. However, this dilutes the concentration of Rb-87 by all the other constituents in the rock. Using average amounts of isotopes:

$$\{[Sr\text{-}87]\,/\,[Sr\text{-}86]\}_t = 22/31 = 0.710$$
$$\{[Rb\text{-}87]\,/\,[Sr\text{-}86]\}_t = 17/31 = 0.548$$

If Rb-87 decay has gone on for 4.5 billion years:

$$e^{\lambda t} - 1 = 0.0671,\ \text{and the intercept}$$

$$[Sr\text{-}87]_0\,/\,[Sr\text{-}86]_0 = 0.710 - 0.0671(0.548),\ \text{or } 0.673.$$

Some references give 0.704.

Comment: This permits a look at the sensitivity of the method. Recall the spread in Rb-87 concentrations in the crust (17±8). Using 0.673, the age calculated for 25 ppm would be 3.1 billion years; for 9 ppm, 8.3 billion years. While there is a large uncertainty in the age, we are still dealing with billions of years, not millions or thousands.

But what about the analysis of meteorites? **SSE**, despite extensive treatment of meteorites, gives only two figures for rubidium content, and these were for total Rb: 7 ppm in certain chondrules in a meteorite, and 2.3 ppm as a mean figure for carbonaceous chondrite type meteorites. Rb is also classed as a volatile element! The earth itself is depleted in Rb vs Sr. Rubidium dioxide melts at 600°C; strontium oxide melts at 2430°C. We should check boiling points.

> **Comment:** It is believed meteorites have been subjected to high temperatures both before and after entering our atmosphere. A loss of rubidium relative to strontium is entirely possible. This would make the calculated ages older especially since we may start with less than 2 ppm of Rb-87.

This information casts doubts on any meteorite ages derived from Rb/Sr dating. We need to see the primary data to estimate the uncertainty in the determination.

Origin of the Earth

Scientists are not agreed on a non-Biblical origin of the Solar System. The Universe is said to be 10 to 20 billion years old. Hypotheses of stellar evolution say our Sun's age is 4.5 billion years. From this and its composition, Science believes it is at least a second generation star. Its age is estimated to be within 100 million years of that of the Earth. So we have three possibilities.

1) Both came into the existence at the same time, though not necessarily by the same mechanism.

> **Comment:** The theories of star formation make this implausible. Creationists agree: the Sun and the Earth had different times of origin.

2) The Earth came into existence before our Sun.

> **Comment:** If Science rejects Creation, it needs to postulate mechanisms for both the formation of the Earth before our Sun and its later capture by the Sun.

> **Comment:** The Bible says the Earth was created on Day 1 and the Sun and Moon on Day 4. The stars were thrown in on the 4th Day, almost as an afterthought. Creationists thus cannot consider our Sun as a second generation star. Interpreting "Days" as long periods of

time conflicts with Science which holds the Earth and the Sun are about the same age.

There is ambiguity about another time period. In **Daniel 9**, we are told "70 weeks" means 490 years. Even so, both camps are agreed on two separate events with their separate causes.

Science has not been able to construct any theory accounting for the Biblical sequence. Of course, God can do anything, but it would seem more reasonable to create the big things first if He is not going to do it simultaneously.

3) The Sun came into existence before the Earth.

   Comment: We will discuss this in Ch. 14 in terms of the origin of the first star. Here we will abstract information from **TDU** and **SSE** on various hypotheses for the formation of the Earth and their flaws. They do not agree exactly but we are interested in what the main facts, arguments and conclusions are. First, some information about our Sun.

Conditions on the Sun

A. Temperature of the Sun

Energy from the Sun impinges on our atmosphere at 0.135 watts/cm^2. By the Stefan-Boltzmann Law for the total radiation from a perfect black body (Exhibit 9.1, Type 4, § C), we can calculate the surface temperature of the Sun. The accepted value for the latter is 5780 K and the emissive power of each unit area of the Sun is:

$$E = (5.670E–12)(5780)^4 = 6329 \text{ watts/cm}^2$$

Since the source of the energy is internal, the core must be considerably hotter to provide the temperature gradient that drives the thermal energy across 432,000 miles, the radius of the Sun, to the surface.

B. Density

The density of the Earth's atmosphere at sea level and 32°F is 0.075 lbs/cu. ft. The gases on the Sun are mainly hydrogen instead of nitrogen/oxygen, the temperature covers a much wider range, but the force of gravity is 28 times that on Earth. Calculations show that the Sun must be a gaseous mass whose density ranges from very much less than atmospheric air on Earth at the Sun's surface to half a billion times that at its core. Further, it is a mass of positively charged bare nuclei (largely protons), free negatively charged electrons, and a certain amount of neutrons.

C. Source of the Sun's heat

    1. Chemical reactions

The energy cannot be from chemical reactions. At ordinary temperatures, hydrogen and oxygen react to form water. But this is reversible. As the temperature rises, the atoms vibrate and oscillate more and more rapidly and finally the bonds rupture to form hydrogen and oxygen ions. These are the corresponding atoms but minus or plus electrons to give them electric charges. Moreover, what is the oxidant? There is not enough oxygen on the Sun, and if water is present, it is not in the abundance expected from centuries of combustion.

    2. Gravitational contraction

Gravitational contraction as postulated by Helmholtz does generate heat but it would have provided less than 0.1% of that actually generated, even if the Sun had started from a very large size (**BDS**).

    3. Nuclear reactions

The only source of the total energy radiated by the Sun over the length of time involved (that Science can think of) is nuclear reactions. Modern thinking is that our Sun works on a reaction with deuterium as an intermediate. We will discuss this in Chapter 14.

Characteristics of our Solar System

Before returning to our examination of our Earth, let us look at some characteristics of the planets in our Solar System. Their orbits are considered ellipses.

    Comment: The eccentricity, e, of an ellipse can be calculated from:

$$e^2 = 1 - b^2/a^2,$$

where a and b are the semimajor and semiminor axes, respectively. If they are equal, as in the case of a circle, the eccentricity is zero. Mercury has the largest eccentricity (0.206), but even this means the b/a ratio is 0.979. For Mars, b/a is 0.996. It is puzzling how Kepler could measure planetary positions closely enough to be able to show that their orbits were slightly elliptical.

Science, in its quest for regularities, came up with the Titius-Bode rule of planet spacings in 1766 which seems like a bit of numerology. We start with a series of integers: 0, 3, 6, 12, etc. Then we add add 4 to each number and divide by 10 to get a series close to the planet spacings in Astronomical Units. The series indicated a planet where the asteroid belt is, and is only good through Uranus. The average deviation from the true spacing is 2.6%.

But let's give our imaginations a little freedom. Let's use a different unit of measurement: the light-year. To avoid very small decimal numbers, use LY/10,000. In Table 12.3, the orbital circumferences are close to simple fractions or multiples of LY/10000. The average deviation is 2.7% and holds through Pluto.

Besides the light-year, this formulation involves the universal constant π. If we do the same calculation for the orbital radii, we find the average deviation is a bit larger: 3.9%. Also the fraction for Jupiter is 13/16 rather than a simple 3/4.

We can also express the planetary masses as fractional multiples of 1/10000 of the Sun's mass, and the orbital velocities as fractional multiples of c, the velocity of light. However, there is no progression or obvious pattern to the sequences obtained. But then, the same was found for the Titius-Bode rule, which is a totally arbitrary formula.

Table 12.3 Orbital Dimensions

Planet	Orbital Circumference		
	cm	LY/10000	
Mercury	3.637E13	0.3844	3/8
Venus	6.798E13	0.7186	3/4
Earth	9.400E13	0.9936	1
Mars	1.432E13	1.514	3/2
Jupiter	4.891E14	5.179	5
Saturn	8.979E14	9.491	10
Uranus	1.806E15	19.09	20
Neptune	2.839E15	29.91	30
Pluto	3.707E15	39.18	40

Comment: If you were assembling a Solar System, you might start with an Earth orbiting a Sun. To cope with cosmic debris, you put in a giant planet like Jupiter to intercept debris coming in from outer space. Then you decide that is not sufficient so you put in three more giant planets: Saturn, Uranus and Neptune. You set different orbital periods so that there is a greater probability of coverage for any incoming direction. You include an asteroid belt to act as a fine screen for rocks that are smaller but that could wreak havoc on Earth. Would you contrive a Titius-Bode rule for deciding on planet spacings or would you use a more direct yardstick like the light-year?

Composition of the Earth

A. Relative to silicon

Table 12.4 gives the distributions of selected elements in three portions of our Universe so you can make your own comparisons. The entries are averages of calculations from data in **EST**, **TDU**, and **SSE**, which generally agree to within about 30%.

Gases are readily lost from small planetary bodies, so the distributions are given relative to a solid element. Silicon is chosen because it is the dominant solid element on Earth. Some significant facts:

- Every element found on Earth is found in the Sun, as you would see if you had the complete table.

- The pattern of relative amounts is quite similar for cosmic rays and the Sun. Beginning with carbon and continuing at least through calcium, they are high for even-numbered elements and low for the odd-numbered elements. This is consistent with building elements by adding helium nuclei since the atomic number increases by 2 each time.

- Since cosmic rays are really matter rays, and have enormous energies, it is reasonable to think they are ejected by stars all over the Universe. This would also explain their random directions.

- The pattern for Earth, if any, seems to be the opposite. It also shows low amounts for the inert gases, nitrogen and carbon.

Table 12.4 Distribution of Chemical Elements

| Element | Atomic Number | Atoms/1000 Silicon Atoms | | |
		Cosmic Rays	Sun	Earth Crust
Gases				
Hydrogen	1	1.4E+06	2.7E+07	1.4E+02
Helium	2	3.1E+05	2.4E+06	6.3E–05
Nitrogen	7	1.9E+03	2.8E+03	1.4E–01
Oxygen	8	7.0E+03	2.0E+04	3.0E+03
Neon	10	1.2E+03	1.5E+03	1.5E–05[a]
Argon	18	8.6E+01	1.5E+02	7.6E–03[a]
Solids				
Carbon	6	8100	9500	1.7
Fluorine	9	120	3.4	3.3
Sodium	11	280	54	120
Magnesium	12	1300	1020	90
Aluminum	13	270	78	310
Silicon	14	1000	1000	1000
Phosphorus	15	56	9.3	30
Sulfur	16	200	480	0.8
Chlorine	17	35	5.3	0.4
Potassium	19	64	3.5	70
Calcium	20	180	60	90
Iron	26	700	820	90

[a]Values are for atoms of gas in the atmosphere/1000 atoms of silicon in the crust.

• Oxygen is the dominant element on Earth. Besides the oxygen in our atmosphere, there is an enormous amount in various solid metal oxides and in water. The basic material of the Earth apparently was (slated to be?) alumino-silicates. It is interesting that magnesium is only 9% of that on the Sun while calcium, a similar element, is 150%.

Calcium goes into bones and shells and magnesium does not. Carbon also is greatly reduced, although it is the basis of all life. Sodium and potassium are increased: these form vital electrolytes in our tissues.

> **Question:** How does this selectivity come about? Did Someone plan a lifeform based on carbon and water that breathed dilute oxygen? If so, why was the more plentiful aluminum not part of our makeup?

The low concentration of the inert gases - helium, neon, and argon - is said to support the argument that the Earth could not have been very hot in the beginning. If it had, then there would have been substantial losses of other gaseous elements and compounds. However, the low concentration can be explained by the low gravitational attraction of a smaller initial Earth, and by the combination of other elements into higher molecular weight substances.

B. Other composition ratios

The Big Bang theory is said to predict the helium content of the Universe at 25% on a mass basis. The figure of 0.089 atoms of helium per atom of hydrogen for the Sun is equivalent to 0.358 g helium/g hydrogen, or 26% of the combined mass. The cosmic ray figure works out to 47% of the mass. So apparently helium is not all locked up in star cores. Instead of choking on a waste product, stars have found a way of getting rid of some of it.

The carbon/nitrogen ratios for the Sun and cosmic rays are about equal. Are they expelled as amine radicals (-C-C-C-C-N-), precursors to amino acids? Our iron/ carbon ratio is 610 times that of the Sun.

> **Comment:** This is just the crust. If the core is really iron, the disparity is even greater. We don't need so much just for our bodies, and it is too hot to have magnetic properties. If the Earth was created, why so much iron? If not, how?

Planet Formation

As to the question of where did the planets come from, let us note these facts about our Solar System:

- It is 24 trillion miles from the nearest star.

- The acceleration of gravity from the Sun is 0.6 cm/s^2 on the Earth; on the nearest star it is a 75 billionth of that.

- All of the planets rotate in the same direction.

Some say these are enough to infer our Sun and its planets most likely had a common <u>local</u> origin.

> **Comment:** Remoteness, if valid as an argument, points to the Sun itself as the source of the System. Yet despite the improbability of a visitor, there could have been one. We also do not know if the planets are all of the same age.

> **Question:** How did the vast empty spaces separating the planets come about?

> **Answer:** "Common local origin" is consistent with a Creator putting the planets in their places in empty space. This is either not good enough for some, or unacceptable to others. After all, are there not Natural Laws by means of which, if we are clever enough, Science could explain what it sees?

Science has offered us three sources of the material from which the planets may have been formed.

- The material was somehow pulled or pushed out of the Sun.

- The material was in space as small particles that aggregated in stepwise fashion.

- The material was in space and swept up by the Sun and the planets as the System moved through space.

What follows is abstracted mainly from **TOG** and **SSE**. The intent is not to present a comprehensive story, but merely to present a flavor of what intellectual paths have been trod. **SSE** takes the following positions:

- Any original nebula probably was not hot.

- Jupiter formed early, probably within the first million years. The inner planets formed later, 10 - 100 million years after the beginning.

- Why the Solar System rotates is unknown. No explanation exists for why nearly 100% of the mass of the System is in the Sun but 98% of the angular momentum is in Jupiter.

A. Nebular hypotheses

After humans got set straight from the erroneous belief of centuries that the Earth was the center of the Universe, and after the Law of Gravity (1687), Immanuel Kant proposed a nebular origin: a cloud of particles of varying sizes moving in all directions. Laplace (1796) conceived of a vast, hot, but highly attenuated nebula rotating around an existing Sun.

The main objection to these is the Ice Skater Objection. Many have seen a skater start spinning and then by pulling his arms close to his body increase his spin speed. This illustrates Conservation of Angular Momentum. Most of the momentum is in Jupiter, whereas it should be in the Sun if the condensation was from a diffuse rotating cloud of gas since essentially all of the mass is in the Sun. Let's check this out.

Let's first calculate the angular momentum for the current Solar System. We focus on the giant planets since the others do not contribute significantly to the total mass or to the angular momentum, J. Calculated momenta are in Table 12.5. I is the moment of inertia of the rotating body; values are found in handbooks.

Calculation: The angular momentum is:

For a rotating solid sphere:

$$J = I \omega = [(2/5)(mR^2)]\, \omega$$

For a mass revolving about an external point:

$$J = I \omega = [mR^2]\, \omega$$

The angular velocity ω, omega, is in radians per unit time. One revolution has 2π radians. Dividing this by the period gives the angular velocity.

Illustration: Jupiter's orbital angular momentum is:

$$J = (1.90E30)(7.78E13)^2\, [2\pi/(3.74E08)] = 1.93E50 \text{ g-cm}^2/\text{s}$$

The other three giant planets do make a significant contribution to the angular momentum. Collectively, though, the Sun has only 3.4% of the angular momentum of the System. The contributions from the rotation of each planet on its axis are negligibly small.

Table 12.5 Angular Momentum in the Solar System

Body	Mass, g	Orbital Radius, cm	Period, s	J, g-cm^2/s
Sun	1.99E33	6.96E10	2.16E06	1.12E49
Jupiter	1.90E30	7.78E13	3.74E08	1.93E50
Saturn	5.68E29	1.43E14	9.30E08	7.85E49
Uranus	8.69E28	2.88E14	2.65E09	1.71E49
Neptune	1.03E29	4.50E14	5.20E09	2.52E49
Subtotal	2.66E30			3.14E50
Total	1.99E33			3.25E50

Next, let's suppose all the mass of our Solar System was once uniformly distributed in a sphere about the Sun's center. Let's assume that it extends out to where Neptune is now, and that its angular velocity was the same as Neptune's now. This should give us a conservatively low figure for the angular momentum because as matter moves closer to the Sun its angular velocity increases. For this assembly we have:

$$J = (2/5)(1.99E33)(4.50E14)^2 [2\pi/(5.20E09)] = 1.95E53 \text{ g-cm}^2/\text{s}$$

So the current Solar System has only 0.167% of the original angular momentum. Where did the rest go?

Suppose we assume only the planetary material was distributed in a sphere.

$$J = (2/5)(2.66E30)(4.50E14)^2 [2\pi/(5.20E09)] = 2.60E50 \text{ g-cm}^2/\text{s}$$

Adding the contribution of the Sun to get the total:

$$J = 2.72E50, \text{ vs the current figure of } 3.25E50.$$

These are close enough to be considered equal, i.e., the angular momentum of the Solar System has remained constant if we assume the planets formed around an existing Sun. As it swept through thousands of light-years of space, it gathered up mass into its nebula from which eventually the planets formed.

Comment: By this reasoning, the Sun came before the planets and they did not have a common local origin. But how did clumps develop in the nebula that grew into planets?

B. Catastrophic hypotheses

Bypassing the question of where our Sun came from, it was postulated that another large body, also of unknown origin, wandered near our Sun and pulled out material from the Sun that became the planets. The body either was another star or a very large mass of stone and iron. This is a tidal theory. It is out of favor because of three main difficulties:

1) The enormous separations between stars make it extremely improbable that two stars would ever get close enough.

2) Any material pulled out of the Sun would probably disperse rather than form planets.

3) The composition of the planets differs from that of the Sun today. The inner planets are stony while the outer ones are gaseous.

Comment: Perhaps the inner planets came from the Sun and the outer ones came from the visitor. Remember that gravitational attraction is a system property: it belongs to <u>both</u> bodies.

Perhaps in an early stage in its development as a star, our Sun was burning lithium, beryllium, or boron and these occurred as runaway reactions leading to something like a Solar belch. And there are Solar flares.

Comment: Have you ever heated a thick oatmeal or a spaghetti sauce? You will recall that steam bubbles form at the bottom of the pot, slowly rise through the mass and burst through the surface, carrying dollops of cereal or sauce with them.

As a star forms by gravitational contraction, its mass gets unimaginably hot. There may be a main nuclear reaction, but this does not rule out side reactions and nonuniform temperature. Finally some of the gases or plasma breaks through the surface into space. Science believes this happens when a supernova explodes, scattering material into space that it <u>cannot</u> recapture. How does Science know? Our Sun contains heavy elements that it could not make, by present theory.

The escape velocity of anything from our Sun is 618 km/s. **TDU** tells us of stellar winds with velocities up to 3000 km/s. The most luminous stars can initiate the ejection of material. For less luminous stars, the mechanism of initiation is unknown, but their radiation pressure can accelerate the ejected material to high velocities. In any case, the rate of mass loss is so great that a large percentage of the initial mass is lost during the stellar lifetime.

Perhaps every star has a planetary system. Or perhaps our Sun was one of a binary system, and the other star exploded. It has even been suggested that there was a ternary system. However, there is no evidence our Sun ever was part of a star cluster.

I do not know if this hot model can be extended to all the other planets. We may note:

- The densities of the four inner planets are about 5 g/cm^3. This speaks of a common mode of formation.

- The densities of the four giant outer planets are about one-third of this. They are low because of large gas atmospheres.

    Question: How were the larger masses flung out farther?

- All planets are said to have cores, but they are not all iron-nickel. Mars' is not. Jupiter and Saturn may have rock cores surrounded by liquid metallic hydrogen above which is liquid hydrogen. Jupiter's core is said to be at 30,000 K. Such temperatures exist just below the surface of the Sun. Uranus may have a rock core.

- Venus and Mars have carbon dioxide atmospheres. Maybe Earth did also, but most of it has been tied up as limestone and dissolved in the oceans. Jupiter is like the Sun: 90% hydrogen, 10% helium. But there is almost total lack of oxygen: the Sun has 740 atoms of O per million hydrogen atoms, while Jupiter has only 1. Saturn has only 6% helium.

C. Accretion hypotheses

There was a return to the nebular theories by revising the mechanism. Little particles grew by colliding with and sticking to other little particles to form planetesimals, which are rocks in a variety of sizes. The rocks then accumulated into planets. Some additional facts tend to support this hypothesis.

- The planets and satellites are extensively cratered, which indicates many collisions with objects in a wide range of sizes in the past.

- Some planets, including Earth, are tilted to the plane of revolution.

Question: What made the rocks cohere? They were probably cold. It is suggested that impact may have caused melting. The Earth's orbital velocity is 18.5 miles/second. A 1-kg rock would be vaporized by its kinetic energy if it struck the Earth with this velocity.

Question: Why don't the rocks in the asteroid belt form a planet? Some say there was a planet, but it was pulled apart by gravitational forces. Others say there never was a planet: the total mass is only 1/3500 of Earth, 1/80th of the Moon, and 1/500th of Pluto. Gravity is thus offered as the "explanation" for both destruction and construction. Perhaps the asteroids represent debris from the Sun that escaped inclusion in the planets.

The choice is between a hot model, in which all the material needed for an Earth is pulled out of the Sun, and a cold model, in which planetesimals gradually collected to an Earth-sized body which then accumulated an atmosphere from space. To compare these, let's look at escape velocities vs the intrinsic molecular velocities of gases.

- The escape velocity from the gravity field of a body of mass m and radius r is:

$$v_e = (2Gm/r)^{0.5}$$

- The root mean square (RMS) molecular velocity, v_s, at temperature T kelvins is given by:

$$v_s = (3RT/M)^{0.5},$$

where R is the gas constant (8.315E07 ergs/g-mol K) and M is the molecular weight. RMS velocities for several gases are in Table 12.6.

Table 12.6 Velocities of Gaseous Atoms

Gas	Atomic Weight	Velocity, mi/s		Radius, miles[1]
		6000 K	3 K	
Hydrogen	1.008	7.57	0.169	130
Helium	4.003	3.80	0.085	66
Nitrogen	14.01	2.03	0.045	35
Oxygen	16.00	1.90	0.042	32
Neon	20.18	1.69	0.038	29
Argon	131.3	0.66	0.015	11

[1]This indicates the radius of the Earth at which the escape velocity equals the atom velocities @ 3 K.

Charles H. Peterson

1. Hot model

The velocity of hydrogen atoms is 7.6 miles/second at 6000 K. The escape velocity of anything from the Sun is 384 mi/s. So the Sun could hold on to nearly all of even its hydrogen atoms.

The escape velocity from the current Earth is 7.0 mi/s. If the initial Earth had the same mass as the current Earth, at 6000 K it should have lost only hydrogen atoms. However, much hydrogen combined to form heavier molecules, like water; oxygen forms many oxides. Most gases would not have been lost from a large, hot Earth.

This is not quite correct. There is a distribution of velocities among the molecules of a gas such that nearly half at a given temperature have speeds greater than the RMS value. So there is a more or less gradual loss of even the heavier gas molecules because the faster particles exceed the escape velocity.

Comment: Why doesn't the loss of molecules cease? The Sun transfers energy to the Earth, heating additional molecules to escape velocity. Over time, the Earth could have lost hydrogen, nitrogen, oxygen, carbon, sulfur, and chlorine in compounds and free radicals as well as inert gases.

2. Cold model

What size Earth should we consider? We can estimate the size from the molecular velocities. Let's treat Earth as a perfect sphere of average density rho, ρ.

$$v_e^2 = 2Gm/r = 2G(V\rho)/r = (2G\rho)(4\pi r^2/3)$$
$$= (5.590E{-}07)(\rho r^2)$$

Even though our Earth now has an average density of 5.5, in the beginning it might have been only stony silicates with a density of say only 3.0. Then:

$$r = (772 \text{ seconds})(v_e)$$

For v_e in miles/s, r is in miles. Calculated values are in Table 12.6. At 0K, all molecular velocities would be zero and every dust particle would hold some gas. Instead, 3 K is used because there is much talk about that being the temperature of background radiation in space. A rock 30 miles in diameter would not hold much of an atmosphere even at 3 K.

Question: How long does it take to aggregate dust particles into a 30-mile diameter sphere? The volume of a 30-mile sphere is 59E18 cm³. The volume of a 1-mm spherical particle is 5.24E–04 cm³, so 112 sextillion particles have to get together.

Where are they? Let us suppose a disc the diameter of Neptune's current orbit (9.008E14 cm) and with a thickness one-tenth the diameter. The volume of the disc is 5.73E43 cm³. If uniformly distributed over the disc volume, there will be about 2E–21 particles/cm³, or 1 particle per 510 trillion cubic meters. They would be about 80 km apart.

Assessment: Accretion from small particles seems not credible. Let's try larger "particles" (planetesimals) the size of the Moon.

Calculation: The Moon mass is 7.35E25 g. In Table 12.5, the planetary mass as 2.66E30 g. Dividing, we find 36,200 Moon-sized bodies are needed to form the planetary system. Each body would have 1.58E39 cm³ of the disc volume. So the <u>average</u> separation is 117 million km. Seems too far. But let's check. The gravitational acceleration between two such bodies is:

$$g \quad = Gm/r^2 = (6.672\text{E}{-}08)(7.35\text{E}25 \text{ g})/(11.7\text{E}12 \text{ cm})^2$$

$$= 3.58\text{E}{-}08 \text{ cm/s}^2.$$

Continued for one Earth year, this acceleration adds 1.13 cm/s to their relative velocity. After 100,000 years, the addition is up to 1 km/s, and the bodies are moving closer by 35 million kilometers a year. Working out the actual time to contact is more complicated, but contact does seem possible in a reasonable time.

Assessment: Accretion of large bodies is credible even though I do not know the process by which large masses were expelled from the Sun. I therefore favor the hot model as the source of the planetary material and the cold model as the mechanism for planet formation.

Conclusion: Overall, it seems the Earth could have come from the Sun, or from another star of similar composition because the mechanism of star formation should be the same everywhere.

Size of the Initial Earth

The initial Earth either was too small to hold an atmosphere or large enough to do so.

- The first option would also have to postulate the accumulation of matter from space. It is speculated that icy meteoroids brought water to the Earth.

Question: How did water in space form meteoroids of pure ice? An enormous number would be required for the 330 million cubic miles of ocean since the volume of a 1.25-mile diameter ball is only 1 cubic mile. The Earth's volume is 260E09 cu. mi.

- In the second option, a large Earth would selectively lose hydrogen and helium. While the Earth was hot, hydrogen and oxygen could coexist as ionized gases. On cooling, they would form water. There would be an oxygen (oxidizing) atmosphere only if much of the hydrogen was lost or tied up chemically. Perhaps the Earth did not have to import water.

Mass of the Original Earth

We can estimate from the compositions of the Earth and the Sun that the mass leaving the Sun might have been 270 times that of the current Earth and, if originating in the outer low density regions of the Sun, occupied a volume equal to that of the present Sun.

Cooling of the Initial Earth

The mass would cool by radiation to space. After recombination of nuclei and electrons into atoms, these would form various oxides, slowly cooling below their boiling points to form liquids, and crystallizing into various minerals. In decreasing abundance, the metals would be magnesium/silicon; iron; aluminum; calcium; sodium/nickel; chromium; and potassium.

The substances with the highest melting points tend to crystallize out first. The actual sequence depends on the relative amounts of the various oxides present, including water. The Earth's crust contains mainly igneous rocks, despite millions of years of weathering. A wide variety of minerals have been found in these rocks, with a corresponding variety of names that is difficult to digest. We can think of these as chemical combinations of alkaline oxides of sodium, potassium, magnesium, calcium and iron with acidic oxides of silicon and aluminum. The most common are:

- Quartz   Pure silicon dioxide

- Feldspar   Alkali aluminum silicates

- Pyroxene   Calcium - (X) - silicate, X being magnesium and/or iron

- Hornblende   Calcium - $(X)_3$ - silicate, X as in pyroxene.

Silicon can tie up a lot of oxygen. The most common igneous rocks are granite and basalt in roughly equal quantities. Granite is a mixture of quartz and feldspar, and seems to be the main constituent of the basement rocks of the continents. Basalt is a mixture of pyroxene and labradorite, which has equal parts of two feldspars (**TOG**).

Temperature History

A. Core temperature

Evidence for a high core temperature begins with the fact that some volcanic lavas are hotter than 2300 K. Then, the temperature in deep mines increases with depth at a rate of 18 - 90°C/km, averaging 30°C/km (**TOG**).

> **Comment:** If the average gradient persisted, a temperature of 6000 K would be reached at only 200 km beneath the surface. So there is a problem with the conceptual "solid" model of the Earth. Perhaps we should view the core as an ultradense fluid.

Iron and nickel melt at 1450°C and 1535°, respectively, at atmospheric pressure. The idea of a solid core again seems inconsistent with this evidence. For reference, of the common oxides, magnesium oxide (MgO) has the highest melting point - 2800°C; calcium oxide (CaO) is lower - 2570°C. SiO_2 (silica) is 1670°C. We should also note carbon, at more than 3500°C.

On cooling, a crust would have formed as a continuous skin on the surface of the magma. We can think of it as the shell of an egg that was repeatedly torn apart by gravitational and rotational forces and melted back into the molten core almost as fast as it formed. Eventually enough of the outer portion of the Earth was cool enough to form a permanent crust floating on the magma. Much later, water would condense from the atmosphere to form the oceans. This would offer a second mechanism, evaporation, for cooling the Earth besides the initial one of radiation.

B. Cooling time

Let us assume that after the Earth left the Sun it quickly formed a sphere of molten material at 6000 K. We wish to know what is the temperature within the Earth as time passes. This is a problem in spherical geometry that is most readily handled by calculus. The result is in Exhibit 12.2. Here we summarize the findings for a sphere of radius a initially at a uniform temperature of 6000 K and exposed suddenly to an external temperature of 0 K. The letter r is the radial distance from the center.

- For r = a, a point on the surface of the Earth, the temperature T is 0 K for all times t > 0 years.

- For r = 0, the center, at time t′= 0, mathematically T oscillates between 12000 and 0 K, averaging 6000 K. This is a peculiarity of the math. For later times, the oscillation disappears.

  o By t′ = 1 trillion years, T drops to 110 K.

  o At t′ = 4.5 billion years, the supposed age of the Earth, T is still 6000 K, or actually whatever the initial temperature was.

- The radial temperature gradient is −11 K/km.

  Comment: Observed values are about −30 K/km. Rather good, considering the assumptions made, but still why the difference? There are heat generating processes, at least within the crust: radioactive decay; gravitational contraction; and frictional heat from movement of tectonic plates. However, these would mean that the core temperature would exceed 6000 K. Prorating, the initial core may have been as hot as 17,000 K.

- The temperature gradient at the center is initially infinite and gradually decreases to zero.

  Comment: We know of nothing in our material world that is infinite. The result is a peculiarity of the math of spherical geometry. It amounts to an infinite capacity to accept or reject heat at the center.

How sensitive are the results to the values used? Let us focus on the temperature gradient at the surface, since we have observed values for this. The geometry is fixed so we can only question our value of z, which depends on material, temperature and time.

- Material

The thermal diffusivity α (alpha) involves three physical properties. For the likely materials, the thermal conductivity k has at least a two-fold range from 0.0022 (clay) to 0.0046 (quartz) for temperatures in the range 0 to 100°C. Iron with elements like carbon, silicon and manganese go to 0.14. The heat capacity c_p has a narrower range (0.18 - 0.22), and is somewhat lower for iron (0.11). The density ρ ranges from 1.8 (clay) to 4.0 (alumina) to 7.0 (iron).

- Temperature

The thermal conductivity increases with temperature, the heat capacity does not change much, but the density decreases with temperature.

- Combined effect

Alpha seems to increase with temperature for the materials of interest. The values of α for oxides range from about 0.0030 to 0.0099 cm/s^2; iron is 0.063. While the core of the Earth may be of iron, most of the Earth is oxides. The alpha used is (0.004134/(0.25)(2.7), or 0.006124 cm^2/s. Table 12.9 shows the results for a few pairs α and t'.

This shows the following:

- The calculation of gradient is not very sensitive to the alpha value over a two-fold range.

- For reasonable α, the only way to get gradients of −30 K/km is with shorter cooling times. It is unlikely that alpha is much less than 0.00500.

Interpretation: A heat source exists either in the crust or just below it. Another possibility is that the Earth is only 600 million years old. Note that in this case, the temperature at the base of the crust is 1067 K, or 794°C, if the 30 K/km persists. A little deeper and the Earth is surely molten.

Table 12.7 Sensitivity of the Gradient Calculation

α cm^2/s	t', yrs	Gradient K/km	Temperature @ 35 km, K
0.00500	4.5E09	−11.8	412
0.01000	4.5E09	− 8.0	281
0.00612	6.0E08	−30.5	1067
0.00612	3.9E09	−11.4	125
0.00612	4.5E09	−10.5	96

Charles H. Peterson

Sources of the Earth's heat

A. Meteorite collisions

Some say the Earth's heat is due to collisions with large meteoroids that impact it as meteorites. This does not seem reasonable. Silicate rocks are good thermal insulators. The heat would not penetrate very far into the Earth and would instead be radiated away.

1. Meteoroid initial temperature

Our scenario begins with the Sun expelling globs of material at 6000 K (or higher) of various sizes. They are too small to hold gases. We ask how long does it take to form a solid shell around the remaining molten material, or arbitrarily, for a shell whose thickness is 5% of the radius of the sphere, when will the temperature at its base be in the range 1500 to 1800°C?

Assume the globs are spherical. By the temperature equation, such a shell would form within 6 minutes for a 1-foot diameter glob vs 10 hours for a 10-foot glob and 300 years for a 1-mile glob. On impact, the shells would break and the molten material within could serve to glue two globs together. These figures support the conclusion that heat transfer into or out of a sphere of silicates is very slow.

2. Meteoroid collisions

We can look at two cases: a) two meteoroids moving in the same direction both at velocity v but pulled together by their mutual gravitational attraction and b) head-on collision. Consider A as stationary so B's velocity is either that due to gravity or 2v. The impact velocity is so low in case a) that negligible heating occurs. In case b), the kinetic energy on impact is shared to raise the temperature of both.

$$KE = (1/2)(m_B)(2\ v)^2 = c_p\ (2\ m_B)(\Delta T)$$

$$\Delta T = v^2 / c_p$$

Thus, masses that have cooled to 0 K can be heated to a melting point of about 2000 K by an impact velocity of 1.14 miles/s. The heat capacity is taken as 0.4 calories/g-K, or 1.674E07 ergs/g-K. Higher velocities mean vaporization and dispersion rather than aggregation of matter.

Head-on collisions might occur often in the early hours or days. Perhaps this is the mechanism whereby the surviving masses all rotate in the same direction.

B. Radioactive decay heat

Measurements in mines show that the temperature of the crust increases about 30°C per kilometer of depth. This temperature gradient is also supported by studies of mineral transformations on exposure to high temperatures. If it continued to the center of the Earth, the temperature there would be 191,000°C, and our model of the core might have to be radically different.

The gradient shows heat is coming up from the interior of at least the crust. If the heat source were in the crust, initially half the heat would be going down. The center would eventually reach the temperature of the source. Thereafter, the temperature of the source would increase to a steady state level.

Is the source heat from radioactive decay? The major contributor to such heat would be uranium-238. In decaying to lead-206 in 15 steps, eight alpha particles and six beta particles are emitted. The fractional mass loss from U-238 is 0.000256, and since this is over 4.5 billion years, the average heat release is 1.22 calories per year per gram of U-238. The crust has about 1.7 ppm of U-238. So decay heat is 2.08E–06 cal/yr-g crust, and the crustal temperature rise is:

$$(2.08E-06) / [0.25 \text{ cal/g-K}] = 8.30E-06 \text{ K/yr,}$$

or 8.3 K per million years, ignoring heat loss. The heat release in a column in the 35-km crust 1 cm^2 in cross-section for a density of 2.7 g/cm^3 is 19.6 cal/ yr. Densities of uranium oxides exceed 7 g/cm^3, but several ores are in the 3 - 5 range, so it would not exist much below the crust. This is in accord with the crustal U/silicon ratio 7 x that in the Sun. So decay does not appear to be a significant source of the heat.

C. Gravitational contraction

Gravitational contraction is a possibility. Using an average density for rocks of 2.7 times water means that the pressure at the bottom of the crust is about:

$$P = (2.7 \times 62.4 \text{ lbs/ft}^3)(22 \text{ miles})(5280)/(144) = 136,000 \text{ psi.}$$

This is more than 9,000 x atmospheric pressure at sea level. At 160 miles, it is over 1,000,000 psi.

Compressing a solid reduces its volume and releases heat, thus increasing its temperature. How much can solids be compressed, and how long does it take to reach equilibrium? A solid may yield only slowly to increasing pressure. Has the Earth's contraction ceased? If so, the heat flow observed is the slow dissipation of previous contraction heat.

Overall, it would appear that the source of the Earth's heat is mainly due to residual heat from that present at its formation or creation.

<u>Summary</u>

1.   Science has no accepted explanation of how our Solar System formed, or for the spacings, sizes, and structures of the planets.

•   Formation of the planets by aggregation from large bodies is credible, but not from dusts.

•   Science cannot explain why the Sun has 99.9% of the mass of System, but only 3.4% of the angular momentum. Jupiter has 60% of the total momentum.

•   Formation of the planets from a nebula around an existing Sun is supported by the constancy of the angular momentum of the System over its life.

2.   Reports of compositions of other planets are based on infrared observations, meaning they represent only the surface compositions.

3.   Science does not know where meteorites (chunks of rock that have been through a fiery reentry into our atmosphere) come from but assumes they are fragments that somehow were knocked out of the asteroid belt.

•   Science assumes meteorites are clues to:

o   the composition of our Earth

o   the structure of our Earth and

o   the composition of our Universe.

• Meteorite ages are estimated from rubidium-87 decay to strontium-87. Rubidium is a relatively volatile element and it is very likely that meteoroids lose a significant part of their rubidium, making them appear older than they really are.

3. Radioisotope decay is the only means for estimating the ages of very old objects on Earth.

• The method requires great care as well as use of judgment in interpreting the observations. The ordinary literature does not disclose the original measurements, so the reader cannot readily assess the uncertainty in the results.

• Nevertheless, it appears that the Earth's age is measured at least in the millions of years, and probably in the billions.

4. Science has only one method for investigating the deep interior of the Earth - earthquake waves.

• The Earth is large enough so that its core is still at its original temperature.

• Heat conduction calculations indicate a central temperature of 6000 K is not inconsistent with a 4.5 billion year age. However, a 600 million year age gives a better match to the observed radial temperature gradient at the surface. The alternative is a central temperature of about 17,000 K.

Exhibit 12.1
Derivation of the Radioactive Decay Law

It is reasonable to assume that the decay rate is proportional to the number of atoms we have at any time. If we had 1000 atoms, 200 atoms might decay in 10 seconds. If we had 2000 atoms, twice as many would decay in that time. We can write this as:

1) $(N_2 - N_1)/(t_2 - t_1) = kN$

where N_1 is the number of atoms at time t_1, and N_2 is the number of atoms at time t_2. Next we need to assume how k behaves over time.

• It is simplest to say k is constant at least over the decay time. This may not be true, but after we have developed a relationship we can test it.

- What about long periods of time, like thousands of years or more? We really don't know. But in the laboratory we observe that k is not affected by chemical or physical tests, with only a few exceptions. The largest known one is only 3.2% for a short-lived artificial (non-natural) isotope of niobium. We can also predict the composition of nuclear fuel that has been in a reactor under simultaneous intense nuclear bombardment, high temperature and pressure for over a year.

- So it is not unreasonable to venture into the unknown with the assumption that k is indeed essentially constant.

What do we use for N? In general, N would have to be at least an arithmetic average:

2) $N_{av} = (N_1 + N_2)/2$

If we consider smaller and smaller intervals of time, we would find fewer decays, but we would approach the instantaneous rate of decay at the time of observation. We can write the left side of the equation as dN/dt, where dN and dt represent very small quantities called differentials. On the right side, N_2 approaches N_1 so N_1 becomes a more and more correct figure for the average number of atoms present.

3) $dN_1/dt = kN_1$, which rearranges to:

4) dN/N = k dt, along with dropping the subscripts. Since N is decreasing, dN is negative. So we insert a minus sign on the right side to come up with a positive number for the rate of decay. It says the fraction decaying in any very short instant of time dt is constant and equal to k.

5) dN/N = - k dt.

From calculus texts, the solution is:

6) $N = N_0 e^{-kt}$

where N_0 is the initial number of atoms, and e is a constant with, unfortunately for convenience, an infinite number of digits beginning 2.71828. . . .

The constant k can now be evaluated in terms of the half-life, $t_{1/2}$, the time required for 50% of the atoms to decay, where ln means natural logarithm:

7) $\ln (N/N_0) = - kt_{1/2} = - \ln 2$

8) $k = (\ln 2)/t_{1/2}$

The half-life is determined by experiment. From it we calculate k. Then from equation 6), we can calculate N for any elapsed time t. The average life is $1/k$.

Putting k in equation 3), we calculate the activity of 1 g of radioisotope:

9) $A = dN/dt = [(1 \text{ g})/(M \text{ g/g-mol})][N_{av}] [(k/31.558E06 \text{ s/yr})]$,

where M = mass of 1 g-mole of the element
and N_{av} = Avogadro Number, 6.02204E23 atoms/g-mol

10) $A = 1.3227E16/[Mt_{1/2}]$ disintegrations/sec per g,

where $t_{1/2}$ = half-life in years.

Effect of uncertainty in k

Equation 7) says that for a given fractional amount of decay, the decay time t is inversely proportional to k. If there is a 1% uncertainty in k, there will be a corresponding uncertainty in t.

Example: The half-life of uranium-238 is 4.47 billion years, making its decay constant:

$k = (\ln 2)/(4.47E09) = 1.55066E{-}10/\text{yr}$

and the decay factor for 4.47 billion years

$N/N_0 = e^{-kt} = 1/2$

If k is actually larger by 1%, the calculated decay time for 50% decay will be 4.426 billion years, or 1% less. If k were too small by a factor of 10, the calculated decay time would still be very large: 447 million years. However, if k were too large, the decay time would be even longer than 4.5 billion years.

Exhibit 12.2
Temperature Distribution in a Uniform Sphere

The temperature, T(r,t), at any point within a sphere due to radial heating or cooling may be calculated from the following equation adapted from **AME**, Article 71, Problem 15. The expression T(r,t) means that the temperature T is a function of (depends on) radial distance, r, and time, t. Calculations will use figures applicable to Earth.

$$T(r,t) = A - 2(B - A) \sum_{n=1}^{\infty} \frac{(-1)^n}{n\pi r / a} e^{-z} \sin(n\pi r / a)$$

A   The uniform surface temperature imposed on the sphere in kelvins.
B   The uniform initial temperature throughout the sphere in kelvins before A takes effect.
a   The radius of the sphere, cm.
r   The radial distance from the center
e   The base of natural logarithms 2.71828. . .
z $= (n\pi/a)^2 \alpha t$, a dimensionless function of time
α   The thermal diffusivity of the Earth, which is k/cρ in cm^2/s.
k   The thermal conductivity, cal/s-cm-K
c   Specific heat, cal/g-K
ρ   Mass density, g/cm^3
t   The time, s, and
n   An integer running from 1 to infinity.

Let's look at this a piece at a time. First, calculate the coefficient of t in z.

α = (0.004134 cal/s-cm-K) / (0.25 cal/g-K)(2.7 g/cm^3) = 0.006124 cm^2/s.

$(\pi/a)^2 \alpha = [\pi/(6.378E08 \text{ cm})]^2 (0.006124 \text{ cm}^2/\text{s})$
        = 1.48582E–19/s

The exponent z is then given by:

z = (1.48582E–19)n^2t, where t is in seconds, or
z = (4.68897E–12)n^2t′, where t′ is in years.

With B = 6000 K and A = 0 K, we have:

$$T(r,t) = 12000\{ \ e^{-(4.689E\text{-}12)t'} \ (\sin \pi r/a)/(\pi r/a)$$

$$- \ e^{-(1.876E\text{-}11)t'} \ (\sin 2\pi r/a)/(2\pi/a)$$

$$+ \ e^{-(4.220E\text{-}11)t'} \ (\sin 3\pi r/a)/(3\pi r/a) - \text{etc.}\}$$

The term $(-1)^n$ provides an alternating series of -1, $+1$, -1, etc.

The temperature calculated for $r = a$ will be that at the surface of the Earth at time t' years. All of the terms $\sin \pi$, $\sin 2\pi$, $\sin 3\pi$, etc. are zero, making $T= 0$. This is as expected since we required that the surface be at the temperature of space, which for simplicity we took as 0 K.

For $r = 0$, the center of the Earth, the calculation is more complicated. First, all of the terms in the braces are of the form $0/0$, which is indeterminate. Dividing the series expansion for the sine by x:

$$\frac{\sin x}{x} = \frac{1}{x}\left[\frac{x}{1!} - \frac{x^3}{3!} + \frac{x^5}{5!} - \cdots \right]$$

$$\frac{\sin x}{x} = 1 - \frac{x^2}{6} + \frac{x^4}{120} - \cdots$$

where x is any angle expressed in radians. The exclamation point indicates a special multiplication operation called the factorial:

$$n! = (n)(n - 1)(n - 2) \ldots 1,$$

so $3! = 3 \times 2 \times 1 = 6$. By trial, we find that for x less than 0.25, the value of $(\sin x)/x$ is within 1% of 1.0.

Now for time $t' = 0$, all the exponential terms are also 1.0, so we have:

$$T(0,0) = 12000 \ \{ \ +1 \ -1 \ +1 \ -1 \ +\ldots\}.$$

For series solutions, one takes the sum of as many terms as are needed for the desired accuracy. Usually the values of the successive terms decrease continually. The meaning here is that $T(0,0)$ fluctuates mathematically between 12000 and zero, averaging 6000 K. Near the center of the Earth, T oscillates to its final value. This, incidentally, shows the mathematics of Science does not always give us an exact, definite answer. However, this peculiarity in this case is only for time $t'= 0$. For later times, the exponential terms eliminate the fluctuations.

Now let's move on to our question: how long does it take to cool from 6000 K? In one trillion years the center will have cooled to 110 K. For 4.5 billion years we get:

$$T(0,t') = 2 \text{ B } \{ e^{-0.02110} - e^{-0.08440} + e^{-0.18990} - e^{-0.33761} + e^{-0.52751} - \ldots \}$$

$$= 2 \text{ B } \{0.9791 - 0.9191 + 0.8270 - 0.7135 + 0.5901 - 0.4678 + 0.3556 - 0.2591 + 0.1810 - 0.1212 \ldots \}$$

$$= \text{B}.$$

The sum in braces proceeds 0.9791, 0.0600, 0.8871, 0.1736, 0.7637, 0.2958, 0.6514, 0.3923, 0.5734, 0.4521, It is clearly converging on some intermediate value. If one plots these numbers against n, the number of each term, one can draw a straight line connecting the odd numbered sums and another one connecting the even numbered sums. These extrapolate to meet at 0.5. By calculation, 18 terms of the series are required to reach 0.500.

The meaning is that even after 4.5 billion years, the temperature at the center is still the initial core temperature, B, whatever value it had.

Temperature gradient

By calculus, we find the temperature gradient, dT/dr, which is how fast does the temperature decrease with distance from the center and with time.

$$dT / dr = -2(B - A)\sum (-1)^n e^{-z} f(r / a)$$

where f(r/a) means a quantity dependent on r/a:

$$f(\frac{r}{a}) = \frac{(n\pi r / a)\cos(n\pi r / a) - \sin(n\pi r / a)}{(n\pi / a)r^2}$$

For r = a:

$$\frac{dT}{dr} = -2\frac{(B - A)}{a}\sum e^{-z}[(-1)^n \cos n\pi]$$

The term in brackets yields the series +1 +1 +1. . ., which adds up to infinity. For time t'= 0, the exponentials are all 1, so mathematically the gradient is infinite at the surface. Physically, the Earth is uniformly at 6000

K, but at the surface, the temperature drops to 0 K in an infinitesimally short distance. However, as time increases, the exponentials eventually reduce the gradient to zero. For the example under discussion,

$$\text{Gradient} = - \, [2(6000 - 0)/(6.378\text{E}08 \text{ cm})][\textstyle\sum e^{-z}] \, [1\text{E}05 \text{ cm/km}]$$

$$= - \, (1.881)(\textstyle\sum e^{-z}) \text{ K/km}$$

The terms involved in the sum are exactly the same as those in the braces in the expression above for the temperature at 4.5 billion years, except that all the signs are positive. Eighteen terms gives us 5.600, and the gradient at the surface is −10.54 K/km.

What is the gradient at the center of the Earth? It involves the sum of terms like e^z/r. For any finite time, the gradient is infinite. For infinite time, we have the question of infinity divided by infinity. Common sense says it should get to zero because the entire sphere eventually is at 0 K. We can reason that for any small nonzero value of r, which makes $1/r$ a large but finite number, we can make t (in z) as large as we like to make the quotient zero.

Chapter 13. Nature of Our Universe

General

In this chapter we turn our attention outward from our Solar System to review what Science tells us about the general characteristics of our Universe, how to estimate its size, what is in it, how it is organized, and how Science answers some elementary questions.

Our Universe is mostly a vast empty space. Yet in the night skies humans have long seen thousands of points of light that they called stars and which Science says are suns similar to ours. Our telescopes still cannot show them as discs, which means they are extremely far.

Space also contains matter distributed as planets, asteroids, comets, meteoroids, dust, molecules, radicals, atoms, protons, and electrons. Everything in the present configuration is moving in various directions with gradual changes in relative positions. Radiation across the entire electromagnetic spectrum is present. However, cosmic rays are high velocity particles of matter.

Spectroscopes reveal that the visible matter in our Universe averages 76% hydrogen and 23% helium, with all the other elements found making up less than 1%. About 100 different molecules and radicals are found (**TDU; SSE**). Most are H_2, CO, OH, and H_2O. Others are combinations of hydrogen, carbon, nitrogen, and oxygen. There is even evidence for glycine, an amino acid. All can be attributed to random collisions between atoms. However, no one has reported finding any life in space.

General Organization of Our Universe

The closest star is about 4 LY (26 trillion miles) from us. Our Solar System is about 28,000 LY from the center of our Milky Way galaxy that has about 400 billion stars arranged in a disc about 100,000 LY in diameter and a central mass about 30,000 LY in diameter.

Photography shows us billions of galaxies, each with 1 million to 10 trillion stars (**EST**, Universe). Not all stars are in galaxies. Our Milky Way galaxy is near the center of a Local Group of some 30 galaxies that has a diameter of about 3 million LY. The nearest galaxies are the Magellanic Clouds, which are estimated at 160,000 to 325,000 LY from us. The midrange is 240,000 LY±35%. The uncertainty of 35% is likely to increase with distance. The next are 3 to 4 million LY farther (**TDU**). The average spacing, however, is 10 million LY (**TFE**). Statistically, the distribution of galaxies is nonrandom. They are in nonuniformly spaced clusters of 10 to 10,000 with up to a quadrillion stars (**TFE**). There are also superclusters.

Some argue that, on a very large scale, the distribution is random (uniform). Yet we see voids several hundred million LY in diameter (**EST**).

> **Question:** Science has yet to conceive of the "natural" processes that produced such organization. Why is there such insistence that the Universe is uniform? Well, it simplifies the mathematical models. Also, Science need not explain the origin and persistence of the nonuniformity we see. Uniformity would mean our galaxy should a) be a sphere of particles or b) not exist at all.

The radius of our Universe remains unknown. We cannot tell where we are relative to its center. Presumably, the Creationists do not dispute the distance estimates. If they also concede the velocity of light (1 LY/yr), the Universe has to be at least hundreds of thousands of years old. This one observation should end the debate about a young Universe. However, it says little about a young Earth.

We may also note another curiosity: Table 13.1 shows the bodies in our Universe are a set of nested sizes. At each level of size, there is a factor of about 100 x 100 between the diameter of the body of interest and that of its containing system. Also the average distance between galaxies is about 10E06 LY vs an average diameter of 1E05 LY, a ratio of 100:1. The Solar System ratio would be 1.70E04 if our Solar System extends to about twice Pluto's orbit.

Interstellar Matter (TDU)

Most of the interstellar matter is in clouds (with space between them), largely as forms of hydrogen.

> **Comment:** How can anyone persist in asserting that the distribution of matter in space is uniform?

The most common type, diffuse clouds (nebulae), are too thin to block light and too cool to glow. They are detected because they redden the light of stars behind them and cause absorption lines in the spectra of those stars. Particle counts are about 1 atom/cm^3 and temperatures are 50 - 150 K. The number density is also given as 0.1 H atom/cm^3. **EST** says 1 atom/cm^3.

Table 13.1 Comparative Sizes of Bodies in Our Universe

Body	Diameter, cm	Ratio
Proton	2.80E–13	
H atom	1.06E–08	3.78E04
Earth	1.27E+09	
Earth orbit	2.99E+13	2.35E04
Sun	1.18E+15	
Solar System	1.39E+11	0.85E04[a]
Our Galaxy	9.26E+22	
Universe	3.75E+27	4.05E04[b]

[a]Pluto's semimajor axis is 3.67E09 miles.
[b]Farthest galaxy, Hydra, at 3960E06 LY.

Dark clouds have masses up to 100,000 solar masses, counts of 1000 to 100,000 particles/cm^3, and temperatures of 20 - 50 K. They contain many types of molecules, including organic molecules. The gas between them is very dilute and very hot, up to 1E06 K.

Question: How can cool regions exist in a hot (high velocity) matrix? What prevents thermal equilibrium? What holds the clouds together? And at 1E06 K, is it a gas or a plasma?

Questions

We will examine the "answers" Science offers us based on its assumptions, hypotheses, and measurements for elementary questions about our Universe such as:

1. Is our Universe infinitely large?
2. How far away are the stars?
3. Is the Universe expanding?
4. What is the age of the Universe?
5. What is the size of the visible Universe?
6. How much matter is there in the Universe?
7. What is the nature of the stars?
8. What is the origin of the stars?

Tools

Here are a few kinds of observations Science makes with its tools in exploring our Universe.

A. Telescope

- Shows directly whether we are looking at a single body, an orbiting system, a cluster, or a galaxy. Coupled with a camera and a photometer yields a measure of apparent surface luminosity.

- Reveals in time the rate of angular displacement of an object. Once the distance to it is known, we calculate its tangential velocity.

- Reveals cyclic variations in luminosity, that are interpreted as either an orbiting companion or an intrinsic pulsation.

B. Spectroscope

- Reveals the identity of chemical elements on the surface of the object being studied.

- Shows relative amounts of those elements by the relative intensity of the lines.

- Shows velocity in the line of sight.

- Shows whether a galaxy is rotating.

C. Hertzsprung-Russell diagram (Exhibit 13.1)

- Organizes brightness/temperature information.

- Permits inferences as to age.

Science is primarily dependent on observations on the radiation from the objects in the Universe for support for its hypotheses. We have learned, however, that our observations may be illusions. An illusion is a mental image that may be a misinterpretation of a real appearance or may be something imagined. Like perspective. Mirages. The velocity of light. What have we overlooked - "dark matter"; interstellar dust; unseen gravity sources; etc.?

Question 1. Infinite Universe

In the search for information and understanding of our Universe, some simple observations have been offered regarding the general size of our Universe. These are summarized here because the questions they sought to answer have not been answered.

A. Infinite Universe, finite amount of matter

Newton apparently believed the Universe was infinite but had a finite amount of matter. Einstein argued against it because the continual loss of light and stars into the void of infinite space never to return was depleting our Universe of matter and energy.

> **Comment:** This is certainly only an opinion, not a quantitative argument. Like the opinion Columbus couldn't sail around the Earth without falling off the edge. How do we know the matter and energy don't return? Isn't Einstein's space curved?

Perhaps the real objection stems from our rejection of the idea of an inevitable extinction of our Universe. But from whence cometh this notion? But if there are no gods, if there is no Creator, if there is no purpose to life, why do we care whether the Universe continues after we are dead?

We have no basis for rejecting the idea of an infinite Universe with a finite amount of matter.

B. Finite Universe

1. Dark night sky

Kepler, Halley, and Olbers argued that if our Universe were infinite, the night sky should be bright because everywhere we looked we should see a star. It is easily proven geometrically if we assume the number of stars per unit volume of space is everywhere the same. The night sky is dark, so either a) the Universe is not infinite or b) the assumption is flawed. Of the various "explanations" under b), two seem to hold up.

• Our Universe is infinite, but relatively young, so that light from afar has not yet reached us.

• Stars have finite lives and die off.

It would seem the dark night sky is an argument for a finite amount of matter in the observable Universe but is not a definitive argument for a finite Universe.

    2.   Finite force of gravity

In an infinite homogeneous universe, the force of gravity should be infinite at every point (Zöllner, 1871: **FG**). It is not, so our Universe is finite. The same rebuttal as for the dark night sky applies: the amount of matter in the Universe is vast but finite, and gravity waves cannot travel faster than light. Let us next review how Science determines distances.

Question 2. Stellar Distances

A. Geometrical Methods

    1.   Nearby distances

By trigonometry, we can find the width of a river without crossing it. We mark off a baseline (L) on one shore and measure the angles A and B between it and the lines of sight from each end to some object on the opposite shore.

$$W, \text{ width} = L / (\text{cotangent } A + \text{cotangent } B)$$

The values of the trigonometric functions are listed in standard tables. This can, in principle, be used to estimate the distance to the Moon. For a 3000-mile baseline, however, the angles are about $89.64°$.

    2.   Large distances

The distance to a nearby star is estimated by parallax. The method uses the same geometry as Eratosthenes used to determine the size of the Earth: if a line crosses two parallel lines, the alternate interior angles on opposite sides of the crossing line are equal.

The parallax of a star is one-half the angle between the lines from the star to each end of the Earth's orbital diameter that is perpendicular to the line between the star and the Sun. It is the only direct measure for stellar distances. It applies only for stars that appear to move relative to much more remote stars as the Earth travels in its orbit. The following is a clarification of the description in **EU**.

Description: The "explanations" typically show a circle for the Earth's orbit. Then lines are drawn to the star at point O from two points A and B that are six months apart in the orbit. These lines make an angle AOB, defined as twice the parallax angle. The line AO is shown exactly in line with a reference star R that is much more remote and thus has no apparent movement as the Earth travels in its orbit. From B, the star R appears in a different position R′ on a line BR′ drawn essentially parallel to AR. Its angular displacement from O is R′BO. Angles R′BO and AOB are equal.

Question: But how does one make this measurement?

- To get the largest apparent angular displacement, the line AB has to be perpendicular to the line of centers of the Sun and O. It would seem that we need to find the time in the year when star O, the Earth and the Sun are lined up. Then A and B are three months before and after this time.

- Star R has to be one that can be clearly identified in its position six months from now. Ideally it will be in the plane of the triangle ABO. It also seems to be asking for a lot to find R exactly lined up with O when the Earth is at A.

For the very small angles involved, the parallax angle p is measured in radians by the ratio of the Earth-Sun distance of 1 Astronomical Unit to OS, the star-Sun distance. The mean value of 1 AU in the following equation was determined by the Jet Propulsion Lab.

$$p = \frac{1AU}{OS} = \frac{14.9597807E12(cm)}{d(cm)}\left(\frac{180°}{\pi}\right)\left(\frac{3600}{°}\right)\left(\frac{\sec}{1}\right)$$

$$= 3.08567627E18/d, \text{ in seconds of arc}$$

d $= (3.2615056/p)$ sidereal light-years.

There is a measurement problem: the parallax angle is at most about 1 second of arc. There are differing claims as to the accuracy of the measurements. **TDU** says 0.01″ of arc with a precision of 0.01″, i.e. an uncertainty of 100%. **EU** says 0.05″ of arc with an uncertainty of only 10%, for which the distance is:

d $= 3.262/0.05 = 65$ LY

AEU says the smallest measurable parallax is 0.1″. A significant part of the reason is that light has to pass through our atmosphere which has a constantly varying density. This spreads the images before they reach our telescopes that add further imperfection.

> **Comment:** Recall how the bending of light past the Sun was offered as support for Relativity on the basis of few hundredths of a millimeter image. One wonders how much the images were spread by our atmosphere and how this was accounted for.

The resolution limit of a telescope mirror is set by the physics of light and the mirror imperfections. For a 10 m mirror, it is 0.012″ (**AEU**).

> **Illustration:** We would see our Sun, 1.4 million km in diameter, as a point of light if it were at least 0.61 LY from us. A more mundane example of 0.05″ of arc is viewing a dime (1.8 cm diameter) at a distance of 46 miles. The star α Herculis has a diameter 800 times that of our Sun and would be seen as a point if only 486 LY away.

3.   Moving cluster method

Over a period of years the radial and tangential velocities of stars moving in clusters are measured to establish the space velocity of the cluster (**TOP**). It is assumed to be the same for all the stars in the cluster and is at some angle to our line of sight. No details are given on how to determine distance from these measurements and the changing appearance of the clusters. This technique can be used for only a couple (2?) of nearby clusters. The Hyades cluster is 150 LY away, with an estimated uncertainty of ±10%. Hyades is of crucial importance because <u>all</u> calibrations of greater distances are based on it.

ACEA gives more description but still does not make the method clear. It says the method is good to 15,000 LY with an average error of 20 to 50%, and has been used on a few hundred stars.

B.   Brightness methods

1.   Energy flux (flow)

A body warmer than 0 K radiates energy in every direction, usually over a range of wavelengths. Planck's Law gives us the intensity of monochromatic (single wavelength) radiation in watts/cm^2-micron.

Integrating over all wavelengths, we get the emissive power, E_s, in watts/cm^2. Multiplying by the surface area we get Q, the total flux.

1) $Q_s = E_s A_s = (\sigma T_s^4)(\pi D_s^2)$

> **Illustration:** For our Sun, E_s was calculated in Ch. 12 as 6329 watts/cm^2. The Sun's area is 6.087E22 cm^2 so its radiant flux is 3.85E26 watts.

The figure in Ch. 12 for the energy received by the Earth is equal to 0.135 W/cm^2. To account for this low figure, think of the radiation from the Sun as passing through the surface of a sphere whose radius is 1 AU. The fraction of radiation received by a 1-cm^2 detector located on this surface would be only $1/[\pi(2 \text{ AU})^2]$.

To apply these ideas to stars, we need to measure the distances to them, and the energy they radiate to us. Unfortunately, the references use several terms about light and radiation differently. **ACEA** is clearer and more specific. Its figure for the Sun's radiant flux matches the figure above.

2. Brightness

Our perceptions of the brightness of stars comes from radiation only in the visible light range. Our eyes cannot detect infrared or ultraviolet light. It is now understood as the equivalent of the sensation produced by a member of a series of neutral grays ranging from white to black. The maximum sensation of brightness is at 556 microns (yellow light). A flux of 1 watt at this wavelength has been defined as a luminous flux of 685 lumens. One watt at other wavelengths will have lower luminosities.

3. Magnitudes

Hipparchus (c190-c125 B.C.) classified the visible stars into six magnitudes of apparent brightness, 1 being the brightest. A second magnitude star was judged only half as bright. The ranking was relative and subjective since "magnitude" was not measured.

> **Comment:** To Hipparchus a red star and a more distant yellow one could appear equally bright even though they emitted the same radiant flux.

His system stated analogously to today's is:

2) $m_1 - m_2 = a \log (B_2/B_1)$,

where the B's are the brightnesses of two stars, a is an arbitrary constant, and the m's are apparent magnitudes. The constant is $1/(\log 2)$, or 3.32.

Question: How did Hipparchus decide to convert a ratio system to a linear system of ranking?

The system was inadequate because for most stars the brightness ratio was not an exact power of 2. Also some stars are brighter than first magnitude, so the scale had to be extended through 0 to negative magnitudes. In 1857, N. R. Pogson modified it to:

3) $m_1 - m_2 = 2.5 \log (I_2/I_1)$

by setting a difference of five magnitudes equivalent to an intensity (apparent brightness) ratio of 100. The ratio for a one magnitude difference is then 2.512 instead of 2.0. The logarithmic form applies because the human eye sees intensity differences as ratios.

Two problems resulted. First, it changed the magnitudes of all the stars. A first magnitude star would now have a brightness 39.8 times that of a fifth magnitude, instead of 16. Second, a zero point on the m scale was needed. **EST** says the zero point was set (arbitrarily) so that most stars retained their usual magnitudes. This, of course, is not generally possible. It means they were changed as little as possible. **ACEA** says the zero was (somehow) set so Polaris would have an apparent magnitude of 2.12.

"Apparent brightness" is the intensity in watts/cm^2 of the luminous flux received from a star. **ACEA** says "luminosity" means the absolute bolometric magnitude and includes all wavelengths. In this book, the terms brightness, luminous flux, and luminosity are reserved for the radiant energy of visible light.

Illustration: ACEA gives some numbers that help understand the equations. A star whose apparent bolometric magnitude is 0.0 has an apparent intensity of 2.27E–12 W/cm^2 on a receiving surface. It is not clear how this was set. Using these values in equation 3) for a reference Star 2 and 0.135 W/cm^2 as apparent intensity of our Sun, the value of m for our Sun is then -26.95. We are still talking about radiation over all wavelengths.

The stars are so distant as to be essentially point sources whose diameters cannot be measured. So we substitute $1/d^2$ for the I's in 3):

4) $m_1 - m_2 = 2.5 \log (d_1/d_2)^2$

This can also be used to convert an apparent magnitude m_2 for a star at d_2 to an "absolute" magnitude M (for m_1), its <u>apparent</u> magnitude if it were brought to an arbitrary distance d_1 of 10 parsecs (32.62 LY).

5) $M = m + 5.0 - 5.0 \log d$

> **Illustration:** For our Sun, m is -26.95 and d is 1 AU, or 4.85E–06 parsecs. Then M = 4.62, the absolute magnitude if the Sun were at 10 parsecs.

> **Comment: ACEA** calls M the absolute bolometric magnitude. This is true only if m was the bolometric measure of the energy received. Otherwise M is an apparent <u>visual</u> magnitude.

There is a complication called extinction. Clouds of solid particles (dust) in the space between stars scatter short wavelength light, thus reducing the amount of blue light passing through them. They also absorb and re-emit light energy at a lower frequency. Both effects make the emergent light redder and fainter, and the distance calculated greater. It reduces m by a quantity A, for which an estimate has been contrived.

> **Comment:** The effect is serious. An extinction of only one magnitude makes the calculated distance 58% greater; two magnitudes, 150% greater. Per **EST**, Interstellar matter, the total extinction is difficult to measure but over a distance of 3200 LY in the plane of our galaxy the reduction in visual brightness is 80%. The distance would be about twice the actual. The relationship is exponential, so doubling the distance in this case increases the extinction to 96%, and the calculated distance by 5.

> **Question:** Has extinction been taken into account in correlating the Hubble redshifts with distance?

Visual magnitudes are measured by combinations of photoelectric cells and filters that respond like the human eye. To measure the total energy emitted, other combinations measure just the infrared or the ultraviolet. When the total flux is measured, the result is the bolometric magnitude.

4.   Hertzsprung-Russell diagram (HRD)

Recall that temperature is our measure of molecular or atomic kinetic energy and involves the concept of zero energy at 0 K. Hence, E_s is an absolute if equation 1) holds, i.e., T and D are the only variables. In practice, we can compare emissions relative to our Sun:

6) $Q/Q_s = (T/T_s)^4 (D/D_s)^2$, where D is diameter.

Taking the logarithm of both sides:

7) $\log Q = 4 \log T + 2 \log D + \log [Q_s/(D_s^2 T_s^4)]$

The last term is a constant calculated from values for our Sun. Knowing any two of the values for another star, we can then calculate the third. A graph of log Q vs log T should yield a straight line for each D. We usually don't know D, or how it affects spectral class, and we're not sure about T. Exhibit 13.1 gives some background on the HRD.

The basic HRD is constructed from stars for which both radiant flux and distance can be measured. For stars down to 15th magnitude, the spectral type is obtained from its spectrogram (**ACEA**). The magnitude can be either the visual M or the bolometric M. It can be the true E only if the diameter is known.

In use, a key fact is that certain types of stars near enough to measure always had about the same luminosity. If a new star was found to be in the same class, it was assumed to have the same luminosity. Its distance was calculated from its M value read from the HRD and its observed m value. To validate, other stars in the same galaxy should be at comparable distances. It is said to be useful to 6000 LY. Its accuracy is low: stars of a given spectral type and class have a wide range of masses (hence, luminosities). So the M value read from the HRD has to be some kind of average for stars of that group.

> **Illustration:** From HRDs in references, the uncertainty in M seems to be at least ±1 unit. What is the uncertainty in the distance d?

$$M \pm 1 = m + 5.0 - 5.0 \log d'$$

Subtracting equation 5):

$$\pm 1 = -5.0 \log (d'/d)$$

and $d'/d = 10^{\pm 0.2} = 1.585$ (+ sign), or
0.631 (− sign)

If d, the true distance, were 100,000 LY it would be calculated as 63,100 LY or 158,500 LY. For ±2 M units, we'd get 40,000 to 250,000 LY.

5. Variable brightness

For more distant objects, Science uses the Cepheid stars. Their average brightness varies directly with the length of the cycle. The distance to the closer Cepheids was obtained by the previous methods.

For any new Cepheid, measuring the period yields its true brightness. We measure its average apparent brightness over its cycle, and then calculate how far away it is by the inverse square law with reference to the Cepheids at known distances. Several Cepheids were found in the Magellanic Clouds. It was assumed they were essentially all at the same distance from us since the diameters of the clouds were small compared to their distance from us.

This technique lets us look out to 16 million LY (**TOP**). **ACEA** says 12 million LY. I do not know what the cumulative uncertainty is.

6. Standard candles

The distance to any farther object whose M is known is calculated from 5) after measuring m. If its spectral type can be identified, its M is found from the HRD. Otherwise, we must make two more assumptions. One, a star cannot be brighter than a certain maximum, which is about $M = -8$. Two, every galaxy should have at least one star at this limit. This takes us out to about 30 million LY (**TDU**).

7. Average brightness

For even more distant objects, the average brightness of an entire galaxy may be used.

> **Comment:** This may be defensible for galaxies since the brightness of the stars in it should be in a range about the average. However, the lay reader has to dig for details. It is clear, though, that the information is getting fuzzy. And Science has no other way to measure vast distances.

Question 3. Expanding Universe

On Earth, we know of the Doppler effect. If a train moves away from us while sounding its horn, we will hear the pitch of its sound drop. The same occurs with all radiation. The whole spectrum of radiation from objects moving away from us will be shifted toward longer wavelengths, i.e. toward the red end of the spectrum in the case of light, while objects moving toward us will show a blue shift.

Spectrograms for various stars show such shifts. The fractional change in wavelength was interpreted as motion away from us or toward us. V. M. Slipher in the 1920s observed large redshifts in 13 distant nebulae. E. P. Hubble, using a larger telescope, found that these redshifts were greater the farther away they were. This seemed to mean velocities increased with distance. Why? Let us look first at how the distance data were obtained and interpreted (**TR**).

A. Distance data

    1.   Distances to nebulae with Cepheids

From distance calibrations for Cepheids, Hubble could determine the distance of nebulae with such stars. It was again assumed that the stars in each were about the same distance from Earth. For the Great Nebula in Andromeda he found a distance of one million LY. Science now says this is two million LY.

    2. Distances to Slipher's nebulae

Hubble used the brightest stars as a measure of distance. Details were not given in the reference, which is disappointing since the redshift calculation was explained clearly. He came up with 20 million LY.

He and M. L. Humason found a cluster of galaxies at one billion LY receding at 60,900 km/s (20% of the velocity of light). Exposure times were 30+ hours. The apparent brightness was obtained by comparison with stars of known brightness using a jiggle camera technique in which the photographic plate was moved around to give a square image of the light on the plate.

> **Comment:** The apparent brightness was really the cumulative amount of light received during the exposure time, not the instantaneous rate of light reception. This assumes at least constancy of the rate. How was the brightness value converted to an average instantaneous rate?

B. Analysis of the data

When the velocities were plotted against the distances, the trend shown was for velocities to increase with distance. The trend was represented by a single line that could be projected back through the origin (the point for 0 distance, 0 velocity). A reproduction of the plot is in **FG**. We note the following:

- Of 31 galaxies, all but two were at least a million LY distant.

- The scatter for most was 50 - 100%: e.g., if the trend line indicated a 500 km/s velocity, the actual velocity was 100 - 800 km/s. This meant considerable uncertainty in the slope of the trend line and in whether one line could represent all the data.

- Of seven stellar bodies within a million LY of Earth, two were moving <u>toward</u> us at velocities up to 250 km/s, and two had zero velocity in the line of sight even though they were a million LY away, i.e., zero <u>relative</u> velocity.

The slope of the line was 500 km/s per megaparsec, and is known as H_0, the Hubble Constant. For reference:

1 megaparsec	= 3.2615E06 LY = 3.0857E24 cm
1 km/s-Mpc	= (1E05 cm/s) / 3.0857E24 cm
	= 3.2408E–20 cm/s per centimeter

Multiplying by 500 gives H_0 = 1.6204E–17 cm/s per cm.

C. Interpretation of the Data

If our Universe is a sphere with a radius R such that it encloses all the objects in it, one interpretation would have been merely that all are moving away from center in different directions.

To explain the trend in speeds with distance, it was postulated that space itself was expanding and carrying everything in it along with its expansion. This, of course, raised all sorts of questions. Several arguments for the expanding Universe were offered.

1. Relativity

This concept leads to equations relating the rate of change of the size of the Universe to the pressure and density of matter in the Universe. We are told that solutions of these equations for various cases all show the Universe is expanding.

Comment: This sounds like the old magic of long-dead priests:

    a.   We have this wonderful knowledge.
    b.   It is reserved to us to know it, to understand it, and to use it.
    c.   There is no way we can explain it to you unless you go through the same difficult path of learning (indoctrination) we did.

Science is not entitled to this position since it denies it to Religion. While non-specialists in mathematics and physics cannot be expected to understand the complex equations involved, this does not make them incapable of understanding the underlying concepts. <u>All</u> equations involve a concept and usually one or more assumptions.

> **Comment:** Einstein patched up his own equations so that they would show a static (non-expanding) Universe. So Science is willing to write equations for opposing "explanations". This gives them an ad hoc character. The equations should fit what Science observes, not what it wishes them to show. Moreover, not all galaxies are receding.

> **Comment:** In one basic regard, there is no difference between a static Universe and an expanding Universe: both ignore the question of origins.

2.   Sole explanation

The redshift data are said to support the expanding Universe concept since there is no other reason why the farthest bodies should be moving the fastest.

> **Comment:** This only means no other reason has been conceived, or if conceived, not accepted. Ignorance is hardly a convincing argument. Have we <u>explained</u> why an orbiting electron, quantized or not, does not lose energy? To say it is quantized only puts a name on the observation and assumes as a matter of faith that it will always be so. Science puts labels on material phenomena it can't explain and accepts them without protest.

Yet when some say there is no possible reasonable explanation for our Universe other than it is the handiwork of God, there is stubborn resistance, rebellion and ridicule. Science labels this position in Religion as fanaticism. Religion should label Science's position as irrationality.

> **Comment:** Suppose we have a spherical container of an explosive in which are embedded metal shot. On detonation, one would expect the

gas released to expand uniformly in all directions. Space itself is not affected. The shot from the outer portion of the sphere would be accelerated by the total expanding gas, whereas interior fragments would not only have less gas behind them but also the expanding gas in front of them might act as a brake. It is not unreasonable for the farthest galaxies to be moving the fastest, if there ever was an explosion.

Comment: For any red shift, the photons in the light don't know we are watching them and obligingly shift to a lower frequency. And they certainly can't accommodate all the potential observers in the Universe. Any shift must be only an apparent shift, different for each observer.

3. Balloon analogy

The distances <u>between</u> stellar objects must be increasing. The analogy offered is that it is like what we see if we inflate a balloon on which spots have been painted. All the spots move away from each other. Several arguments can be offered in rebuttal.

a) The analogy is deceptive and glib.

Certainly there are individual stars in the Universe. But there are also huge galaxies. There is a great deal of space between its member stars. Why doesn't <u>this</u> space expand? Why are there still structures like spiral arms on the periphery of galaxies despite their supposed vast age?

Instead of spots let's paint clusters of spots on the balloon. Then when we inflate it, we see clearly that:

• The spaces <u>between</u> the dots in the clusters also expand; and

• The dots themselves expand. The promoters of the analogy don't call our attention to this.

It is argued that gravity holds the galactic structures together.

Comment: This requires the space within galaxies to be different from that outside the galaxies. Which is remarkable, since we have been unable to identify any properties for space in the first place. It also requires the same effect on space from galaxies of vastly different masses.

b) If there ever was a Big Bang, the motion of the stellar bodies is actually radial from a center that is not necessarily Earth.

The spots are really moving farther from the <u>center</u> of the balloon and that is why they seem to move farther from each other laterally. We can see this readily in two dimensions.

Illustration: From a point O on a piece of paper draw a vertical and a horizontal line. Show the Earth at an arbitrary point E on the vertical line moving away from O and a galaxy at point G anywhere to the right of E and closer to the horizontal line and moving away from O. Draw the line of sight EG.

Now show the situation some time later when E is at E′ and G is at G′, both farther from O. E′G′ is clearly larger than EG. Radial divergence is thus a sufficient explanation for the observation that G is moving away from us.

c) We have no information on the other components of velocity.

Comment: In general a star travels at an angle to our line of sight. We cannot see the angle, but sometimes we can calculate it. We have the radial component of velocity from the Doppler effect (redshift). We need the tangential component and the third component perpendicular to the plane of the radial and the tangential components. Then we combine them by means of the Pythagorean theorem.

Question: How fast are these components changing? This is unknown. The enormous distances mean an angular displacement would be detectable only after centuries.

d) We cannot evaluate the Hubble Constant for nearby galaxies.

Nearby galaxies, including the Andromeda galaxy two million LY from us, have random motions that are much larger than their motions due to the expansion of the Universe (**BHQU**). These are said to overwhelm those calculated from the redshifts.

Comment: I really don't understand this argument. Regardless of the size of the other two components of velocity, the one in the line of sight should be measurable by the red shift, especially for nearby galaxies.

Andromeda rotates, giving stars in the spiral arms a velocity of 250 miles/s. Multiplying the Hubble Constant from § E, below, by the distance gives us 67 miles/s which should be the galactic recession velocity. Surely this can be detected.

4. Cake analogy

It is argued that the situation is like that of a cake with raisins. As the cake rises on baking, it drags the raisins along with it.

Comment: If valid, it implies that all structures in the Universe are embedded in an undetectible matrix. Science has abandoned an ether for conducting light. Has it been resurrected for a different phenomenon and not yet given a name?

Question: How can space itself expand?

The cake expands because gas bubbles expand when heated, and chemical reactions of baking powder generate more gas molecules. More cake does not appear to fill the space where it was. Nor is more space created. In other expansions not involving gas bubbles, the expanding medium is reduced in density and it develops holes.

If the expanding space will drag the galaxies along with it, either this space is capable of indefinite expansion or more space has to appear. Science has created another mystery. We started out to explain the origin of the Universe and now we also have to explain where new space is coming from. This is in direct conflict with Ockham's Law.

Question: Who or what is stretching space? Or, since we inflate a balloon by putting a gas into it, who or what is creating more space? It is amazing how many are willing to believe that a nothing has properties. But such are not consistent in their logic. They say God does not exist, i.e., He is a nothing, and are unwilling to give Him properties.

Question: What is space expanding into?

GNP argues that the Sun can stretch space by its huge gravity. Thus the Universe expands without having to expand into some external void.

Comment: Conceptually incomprehensible. Does space expand only if there is some large mass causing it? Gravity is only attractive. If the Sun is pulling on space locally, the only way it can be stretched is if space has some structure and the other end of it is attached to something, or is being pulled by some other mass.

Question: Let us remember the assumed Cosmological Principle that the laws that govern our immediate world apply everywhere in the Universe. When we stretch something on Earth, the material between the two sources of the stretching is redistributed so that at the halfway point there is less material. Does this not apply to the stretching of space?

Comment: What we experience, always, is that when something expands, something else has to give way around that something. Otherwise we are talking the same kind of mystical nonsense of which Religion is accused. Many basic objections to Religion rest on the claim that we should believe only what we can experience in the "real" world. Why is this not required of "scientific" explanations?

D. Further critique of the expanding Universe

1.   Gravity has a long-term effect on velocity.

The light from the distant structures really is from the earliest moments. The light from the nearer structures suggests that something, perhaps gravity, has operated to slow them down. This could be enough to explain the pattern of velocities.

2.   There may be other causes of redshifts.

One astronomer, William Tifft, observed that spiral galaxies often had higher redshifts than elliptical galaxies in the same cluster. So the total redshift cannot be due to just the recession velocity. He claimed they change over time and postulates that they are quantized. If true, the Universe is not expanding, and the Big Bang may not have been (Man Stops Universe, Maybe. D. Sobel. **Discover** - April 1993).

Comment: However, this does not rule out something like a Big Bang to explain why such a large percentage of stellar bodies appear to be moving away from us through space or to predict a future reversal of the direction of motion. The expansion of space itself is thus an irrelevant and unnecessary concept that so far is impossible to explain.

3.   There is motion in other directions.

As noted earlier, a streaming of galaxies toward a region called the Great Attractor in the constellation Centaurus has been observed. This is incompatible with everything fleeing outward from some center.

4.   The velocity changes are ancient.

The velocities obtained are those of millions of years ago. They tell us nothing of what they are today. If a thousand years from now we could measure them again, we might estimate how fast they are changing.

> **Comment:** Consider what a tiny fractional increase it is: only 18.3% every billion years. (**BHT** says 5 - 10% per billion years.) However, if the motions were randomly distributed, some of the radial velocities should be toward us, so it is indeed noteworthy that the distant ones all seem to move away.

5.   The equations are for matter, not space.

The equations involve the following variables:

R   The radial distance of an arbitrary point from O, the site of the Big Bang.

dR/dt   The velocity of the point away from O.

d^2R/dt^2 The acceleration of the point away from O.

ρ   The mass density of the Universe, rho.

p   The pressure in the Universe.

These imply modelling the Universe as a sphere of radius R and average uniform density of ρ (rho). These are attributes of the matter <ins>in</ins> the Universe. Space itself has no mass, density or pressure. Neither does the space in an empty cylinder of gas.

> **Comment:** Be aware that such equations say nothing about where point O is relative to where we are in the Universe. Later when we talk about the size of our Universe we mean the radial distance from us.

Points are a convenient mathematical fiction. In practice, we would be talking about how far one body is from another as the only way we can talk meaningfully about size changes. Space itself is homogeneous; it has no milestones. There is absolutely no way of measuring the stretching of space except by reference to the positions of matter in it.

In general, ρ varies with distance. The equations are complicated enough so it is assumed there is an average density that applies everywhere in the Universe. This may be reasonable for some purposes, especially since the density is so extremely low. However, it is figured on the quantity of observable <u>matter</u>. Then the average spacing is used to associate a volume of space to that quantity of matter. The ratio of matter to volume gives us the density. But it is not the density of space: that is zero. It is the amount of matter per unit volume of space. If "dark matter" were found, that would be something <u>in</u> space. A similar consideration applies to pressure.

    6.   Another inference is reasonable.

Let's take one more step. Whether or not the Universe is expanding, the evidence is that most galaxies seem to be moving away from <u>us</u>. A logical inference is that, if there was a Big Bang, it had to be right where we are because that is where all these radial velocities are said to project back to.

You know, sort of like Creation. . . .

E.   Critique of the data

This concludes our review of how the data were obtained and interpreted. Let's now take a closer look at the original data on four distant galaxies obtained by a later worker Sandage as shown in Table 13.2.

Table 13.2 Apparent Expansion of the Universe

Galaxy	Distance	Recession Velocity,	Y,
	LY	cm/s	cm/cm-yr
Virgo	22E06	1.2 E08	1.819E–10
Corona	400E06	2.15E09	1.793E–10
Bootes	700E06	3.93E09	1.873E–10
Hydra	1100E06	6.09E09	1.847E–10

Virgo is four times farther than the most distant one Hubble studied. Dividing velocity by distance gives the fractional increase in distance per unit time, which is H_0, the Hubble Constant for the galaxy. Multiplying by 31.558E06 s/yr gives Y, the fractional increase/year. If we plot Y vs distance, we get a straight line with essentially no slope by visual inspection. By the standard mathematical technique of Least Squares:

$$Y = (1.810E-10) + (4.389E-39)R, \text{ where R is in cm.}$$

Comment: Least Squares is designed to minimize the sum of the squares of the deviations of the actual data from the correlating line. It is not the only criterion for how to fit a line to data, but it is the one most commonly used.

Comment: The approach is empirical: the Big Bang is only a concept and has no theory to provide a basis for a quantitative relation between velocity and R. Even for Hydra, the farthest galaxy studied, we see the contribution of distance to Y is:

$$(4.389E-39)(1100E06)(9.461E17 \text{ cm/LY}) = 4.568E-12,$$

or only 2.5%. Still using standard statistical techniques (**QCIS**), we find that the uncertainty in Y is about 10%, so we are not justified in concluding there is a real increase in H_0 with distance (or hence with time). Meaning, the fractional increase of the oldest, and hence the earliest, is statistically no greater than that of the youngest. Each galaxy recedes at a constant percentage each year. There is no tendency to slow down with distance to 1100E06 LY and time to 1100 million years.

Comment: AEU asserts the Hubble parameter, H_0 which is dR/Rdt, decreases with time. The data don't agree.

The constant recession percentage is said to hold for the galaxies Hubble studied. Except for those moving toward us. Also Sandage reported finding six galaxies farther out that appear to be travelling faster than indicated by the Hubble plot. These data will be discussed further in the section on Singularity in Ch. 14.

Question 4. Age of the Universe

Our Universe appears to be very old, based on information from several fields of investigation. It is also believed that light from the stars has taken at least millions of Earth years to reach us. If the light was from an exploding star, and if Science is correct on the life history of a star, that star must also have had a long life time prior to the explosion.

The HRD gives us another clue. Science has postulated a mechanism for the Sun's energy output that means the Sun's life is about 10 billion years, about half of which is past. Stars at the top of the main sequence are more massive, hence are more luminous, burning their hydrogen faster, and therefore are younger. Stars lower down on the main sequence are older since they burn hydrogen more slowly. It is asserted that the diagram is calibrated so that position shows the age of the stars. Details of the calibration are not given in popular texts.

The reciprocal of H_0 is an estimate of the age of the Universe as $1/(500$ km/s per Mpc), or 1.96 billion years. Unfortunately, this conflicted with the 15 - 20 billion year ages derived from models of star evolution, and with geological estimates of the age of the Earth as 4.5 billion years.

Then it was found that there were two kinds of Cepheid variables, so the distances were increased but it is not clear by how much (**TDU**). **FG** says the factor went to 3 because a) there were at least three kinds of Cepheids, b) the statistical methods for extracting the distances to the Cepheids in our galaxy were revised, and c) the effect of interstellar dust (on the red shift) was discovered. Other corrections were for errors in earlier values of apparent magnitudes, the actual brightnesses of the brightest stars in large galaxies, and erroneous identification of hot nebulae as brightest stars.

> **Comment:** This is characteristic of Science. However, without the data from star evolution and the age of the Earth, Science would have had no reason to look further. Religion has also changed some of its teachings about our Universe.

Sandage's later work yielded a much lower figure: 55 km/s per Mpc, increasing the age of the Universe to 17.9 billion years. Some support 50; others have found values in the range 75 - 100 km/s per Mpc.

Question 5. Size of the Visible Universe

The Hubble Constant can be used in reverse. Given an observed velocity for a stellar body, v, its distance can be calculated as v/H_0. However, the range of H_0 values means an uncertainty of as much as ±43%.

A. Estimates of the radius

The Sandage work showed the radius of the Universe has to be at least 1.1 billion LY, if we are its center.

• In **CREA** we find 15 billion LY, but it is also noted that two galaxies were reported at 17 billion LY. These figures are estimates of the radius if there is some other object a comparable distance from us in the opposite direction.

• **AEU** says that beyond 30 LY the distance measurements become increasingly poor and are often uncertain by a factor of 2 or 3.

B. Estimates of the Hubble radius

H_0 is now in the range 40 - 100 km/s-Mpc. Because of the unresolved uncertainty in its value, **TEU** introduces a parameter h, which is a dimensionless factor in the range 0.4 - 1.0. Then $H_0 = 100$ h $= (3.24078E{-}18)$ (h) cm/s per cm. The reciprocal of H_0 is interpreted as the time the light from distant stars has been travelling to us.

$1/H_0$ = Age of Universe = (3.086E17)/h seconds
      = (9.778/h) billion years

The age then is 24.4 to 9.78 billion years. The distance light has travelled in this time is the Hubble radius, R, and is 24.4 to 9.78 billion LY.

C. Initial size of the Universe

The average H_0 is equivalent to 2.268E–18 cm/s per cm. The fractional increase in distance is then 7.159% per billion years. Looking backward in time, the Universe would be contracting at this same rate. Many have inferred the Universe started as a point. If this rate was constant, the differential equation for the way the radius of the Universe decreases going back from the present is:

$dR/R = -k\ dt,$

where dR/R = the fractional change in R over a small time interval dt

      k = 0.07159/billion years
and   t = time in billions of years. Integrating:

$R/R_0 = e^{-kt}$, where R_0 is the Universe radius today

For t = 14 (billion years), we find R = 0.368 R_0, or about one-third of what it is today. So talking about dimensions like 1E–33 cm at the time of the Big Bang is totally unjustifiable.

Comment: Some argue the rate of expansion in the early ages was much faster than it is now. How is this known? The Universe thus would be younger and initially larger than 0.369 of today's size. How fast did the rate decrease over time?

Summarizing, we have these estimates for R:

CREA 15 - 17 billion LY
TEU   10 - 24 billion LY
AEU   2 to 3 x all remote distance estimates

The smallest is 2 x 10, or 20 billion LY, and the largest is 3 x 24, or 72 billion LY.

Question 6. Total Matter in our Universe

A. Estimates of the radius of our Universe

Estimates of the size of the Universe have increased dramatically over time. We need to associate them with estimates of mass and density. In doing this, we must consider mass that is in stars and mass that is between stars, i.e., interstellar matter.

Illustration 1: Our Solar System

Our nearest neighbor is α Centauri: 4.34 LY, or 4.106E18 cm, away. Using this as the average spacing for stars within our galaxy, the spherical volume allotted to our Solar System is 3.62E55 cm^3. Almost all of the mass in our System is in our Sun: 1.99E33 g. Spread out uniformly, the average density of our System would be 5.49E–23 g/cm^3, corresponding to about 33 protons per cm^3.

Illustration 2: Galaxies

We are told the nearest galaxies to us are about 75,000 LY away (**EST**, External galaxy). However, in the same article we are told the nearest galaxies to ours, the Milky Way galaxy, are the Magellanic Clouds at about 150,000 LY.

Also, a typical galaxy has a mass of about 2E11 times that of our Sun. Using the larger figure for distance, the volume of our galaxy is 1.50E69 cm^3, and the mass is 3.98E44 g. The density is 2.65E–25 g/cm^3, corresponding to 0.159 protons/cm^3.

AEU says there is about 1 galaxy per 1000 cubic megalightyears. This is a separation of 12.4 million LY vs the 150,000 LY figure above. So

271

we have another discrepancy to stumble over in seeking understanding. These examples show we have a problem in finding values for the numbers we need.

B. Estimates of density of interstellar matter

The range is wide for clouds: 1 to 100,000 molecules per cubic centimeter. And we don't know the relative amounts of these clouds. It is a serious problem because there is supposed to be more matter in interstellar space than in all stars. In cloudless regions, particle densities, except for an **SSE** estimate, appear to be far less than 1 per cubic centimeter.

C. Estimates of total matter

One also finds in the literature a wide range of estimates of total mass, density, and number of particles in our Universe. Telescopes of Hubble's day could see 300 million LY into space. Around 1950, the radius was thought to be 2 billion LY. Now the figure is in the tens of billions of LY. As for matter, we have:

- Sheldon (**STR**): Distributing all the matter in the Universe evenly would make the density 3E–25 g/cm^3.

- Gamow (**BDS**): Distributing all the matter in the stars uniformly throughout space would make the average density 1E–22 g/cm^3. Gravity would break the mass into Sun-size spheres that would eventually contract into stars.

- Dirac (**FG**): There are 1E80 baryons (protons plus neutrons). The statement is dated about 1937.

- Hoyle (**TNU**): The background material in space is about 1 atom/pint.

- Krauss gives two estimates in **TFE**. There are on the average about 100 billion stars like our Sun in each galaxy and there are about 100 billion visible galaxies. The total mass is $(100E09)^2$ x 1.989E33, or 1.989E55 g.

Based on luminosity, ρ_{av}, the average density of the visible matter, is 4.5h billion solar masses per cubic megaparsec, where h is the uncertainty factor on the Hubble Constant. For a Universe radius of 10E06 LY, the mass is 7.55E53 g. This is smaller than the previous estimate by a factor of 22.

$$\rho_{av} = \frac{(4.5E09)[1.989E33(g)](h)}{[(1E06(pc)(3.086E18(cm/pc))]^3}$$

$$= (3.046E-31)h \ g/cm^3$$

$$= 2.132E-31 \ g/cm^3 \text{ for } h = 0.70.$$

In the above, pc means parsec.

- Hawking (**BHT**): The observable Universe has 1E85 particles. This apparently includes protons, electrons, neutrons, photons, neutrinos, etc.

We hear of "dark matter", something that is assumed to exert a gravitational force like ordinary matter, but is undetectible because it has no charge or anything that is measurable by instruments.

Comment: It is, however, incredibly inconsistent and confuses the meaning of "matter". Gravity is pervasive and is readily detectible. Isn't it a space warp? Consider also that this "dark matter" is not just "out there", but is everywhere, and may be in us. Orbits of planets have been calculated with high precision without considering dark matter.

Comment: Science postulates an intangible "dark matter" in its effort to explain the unexplained. Yet it scoffs at tangible events that Religion attributes to spirits and angels. How many ideas has Science scoffed at that later were validated? And how many ideas has it offered that later were proven wrong? What evidence will convince Science that there is a world besides the material?

D. Critical density

The mathematical descriptions of our Universe lead to three types: a closed curved one, an open curved one, and a "flat" one. In the first, eventually gravity will cause it to collapse back on itself, perhaps cyclically. The second will expand forever. The third will also expand forever but at a decreasing rate so that it approaches a fixed size. The meaning of "flat" is not clear, unless it means that space itself is not curved, whatever that means.

Question: If the Universe is not curved but "flat", what becomes of Einstein's explanation of gravity as a curvature of space? How much is the curvature anyway? If it is due to mass, then space must be very bumpy from the various size masses out there. What is their effect on the path of a light ray?

The choice depends on the average density of matter in our Universe. The density for a flat Universe is the critical density, ρ_c. Apparently, scientists find a "flat" Universe psychologically more acceptable than an open or closed Universe and so show their bias by believing that the actual density is very close to the critical density even though the observable matter is only about 1% of that for the critical density.

- **TEU:** A formula for the critical density below which the Universe will expand forever is:

$$\rho_c = 3H_o{}^2 / 8\pi G = 1.8791 \ h^2 \ E-29 \ g/cm^3$$

For $h = 0.70$, $\rho_c = 9.2076E-30 \ g/cm^3$.

- **TFE:** The observable baryonic matter (like protons) can account for no more than about 12% of the total density. So indirectly we find that the observable Universe protonic content has to be no more than $1.10E-30 \ g/cm^3$ ($h = 0.70$) if the Universe is "flat".

- **BHQU:** The closure density, which presumably means the same thing, is $4.7E-30 \ g/cm^3$. This would correspond to an h of 0.50.

- **TDU:** The critical density is $6E-29 \ g/cm^3$, based on a Hubble Constant of 55 km/s per Mpc.

- **EST**, Galaxies: The critical density is $2E-29 \ g/cm^3$. This would correspond to an h of 1.032.

There is a factor of 60 between the lowest and the highest values for critical density. Only **TEU**'s estimate of actual density is close to the critical density. All others are greater, meaning that they lead to a closed Universe. With the uncertainties in h and the closure density, we have no objective basis for selecting a Universe mass.

For the purposes of this book, I chose $h = 0.70$ which corresponds to a closure density of $9.20E-30 \ g/cm^3$. Combining with the Universe radius of 14.0 billion LY gives a Universe (spherical) volume of $9.732E84 \ cm^3$, and a mass of $8.96E55 \ g$. If all the mass is present as protons, we get 5.5/cubic meter and a total of $5.36E79$ protons.

Calculation: Let's check on an earlier question: how much matter is in galaxies vs that in space?

Using a particle density of $1000/m^3$ for interstellar matter, the mass density would be $1.673E-27$ g/cm^3 if they are all protons. There would be more matter in interstellar space than in all the galaxies. Only if there were less than 1 particle per m^3 average would we get an average mass density close to the critical density.

Questions 7, 8

We will address these in the next chapter.

Summary

1. Our knowledge of our Universe is based almost entirely on study of the radiation we receive from it. The rest is from the few fragments coming to us as meteorites whose origin we can only guess at.

    •   The best guess is that there is a vast but finite amount of matter in a variety of sizes and shapes in an effectively infinite Universe.

    •   Estimates of its size, mass, density, and light/mass ratios cover wide ranges.

2. An average picture is a Universe of visible matter spread over a radius of 14 billion LY to a density of $9.2E-30$ g/cm^3. The real Universe may be much larger. We have no way of knowing where we are in this Universe, or of knowing its actual radius, if it has one, except in terms of how far matter is from us.

    •   The farthest direct measurement of distance is 65 LY, with an uncertainty of 10%. It is made by parallax measurements. Distances beyond 30 LY may be uncertain by a factor of 2 or 3. Such measurements involve assumptions about true brightness.

3. Space contains many clouds of matter that are too thin to block light and too cool to glow that are detected since they cause light from distant stars to undergo a redshift.

    •   Science postulates that temperature is due only to molecular motion and then confuses the issue by talking of the "temperature" of empty space.

- Many kinds of molecules and radicals are found in space. But no life of any kind.

- Science cannot explain the organization exhibited by the matter in the Universe.

4. Stars have been grouped into spectral classes in which the surface temperatures are equal and constant. This permits an estimate of the absolute magnitude for a newly found star and coupled with a measured apparent brightness yields its distance.

- It is believed that the period of a Cepheid indicates its true brightness, from which its distance can be obtained as above.

- It is assumed a star's brightness is limited and every galaxy has a star of this maximum brightness (if it can be found among the billions of others).

- For remote galaxies, the average brightness of the galaxy is used, but the method is not clear. It may involve comparison with the brightness of stars of known distance.

5. Science offers us bizarre "explanations".

- Photons

Particles that travel at the speed of light, have an unspecified size, have no mass or charge yet have momentum and frequencies but no amplitudes, and interact only with matter, but two beams of radiation can cancel each other out.

- "Dark matter"

Something that is 100x the mass of the total visible matter, exerts a huge gravitational force on the visible matter, but cannot be detected.

- Curved space

In three versions. What is actually curved? Gravity is said to be due to the curvature of space, yet the density of matter in it is supposed to be very close to the critical value at which the Universe is "flat".

- Expanding Universe

The Universe, and especially space, is expanding. The concept is based on the sole evidence of redshifts of stellar radiation.

- Virtual particles

Matter that appears unpredictably out of the vacuum of space and disappears so quickly that its energy cannot be measured.

6. There are numerous objections to the assumption of an expanding Universe and its rationalizations.

7. The concept of the Universe beginning as a single point without dimensions is contradictory, untenable, and probably ridiculous.

Exhibit 13.1
Hertzsprung-Russell Diagram

Classification of stars

A spectrogram is a photograph of the light from an object and consists of a series of lines corresponding to various electronic transitions in the atoms of that object. The wavelengths of the lines are shortest on the left and increase toward the right. The lines may be so close as to appear as a continuous band of color.

Stars are made of hydrogen and helium; less than 2% of their mass is due to other elements that add other lines to the spectrograms. Similarities permit grouping stars into spectral types according to the strongest lines. Each type is given an identifying letter A, B, O, etc. (Was there ever a C, D, E, H, etc.?)

If the spectrograms are placed in the order O, B, A, F, G, K, M, the patterns formed by the lines are progressively displaced to the right. Now the "color" of starlight ranges from infrared to ultraviolet. This progression was observed in heating metals, so it was inferred that it also applied to stars. **TOP** states that brightness, surface temperature, and color depend only on the mass of the star. Temperature is thus involved in both luminosity and spectral type.

Hertzsprung-Russell diagram (HRD)

The HRD was the result of independent work by E. E. Hertzsprung and H. Russell on relating luminosity and spectral type. M (visual magnitude) was plotted in the vertical direction vs. spectral type in the horizontal, both on linear scales. The points lay along a broad diagonal band from upper left to lower right, with many other points randomly placed. The band is the main sequence and contains 99% of all stars in classes B to M. (Cool) red dwarfs are at the lower right; (hot) blue giants are at the upper left.

The scatter was large: luminosities of Type G stars like our Sun spanned a range of 10 billion. Modern plots are said to show a range of only 100. Many stars off the main sequence are now shown lying on six irregularly spaced more or less horizontal bands called luminosity classes. The five on the right of the main sequence are various size giant stars. On the left below the main sequence are white dwarfs. Many stars still do not fall on these lines.

The surface temperatures shown for spectral types are in Table 13.3. The sequence for O, B, A, and K is roughly logarithmic (**TDU**, Fig. 20.5). Also, both ends of the band show curvature. So whatever variables are involved, it is not purely log T. If it were, the plot would be a single straight line at a 45° angle.

TDU says M, the absolute magnitude, is used for log L since they are linearly related, provided M is the bolmetric magnitude, meaning the measurement of apparent brightness m includes infrared and ultraviolet radiation. So there is a bit of confusion here for the reader about the meaning of M.

Table 13.3 Temperatures of Spectral Classes

Spectral Class	Temperature, K
O	50,000
B	20,000
A	10,000
F	8,000
G	6,000
K	5,000
M	4,000

Age of star clusters

Based on concepts of stellar nuclear reactions, the life of our Sun is estimated as 10 billion years and it is halfway through that life. It is assumed that all the stars in a clusters are about the same age. An HRD is prepared for the stars in a given cluster and compared to the standard HRD, but the descriptions I have found thus far do not make clear how the time scale is established.

Chapter 14. Explanations for our Universe

Possibilities

For decades astronomers and physicists have labored to interpret their numerous observations about our Universe. They have discarded many earlier hypotheses. We seek to identify the facts and assumptions behind some of the surviving hypotheses. Our Universe is mostly hydrogen, so it is reasonable for scientists to wonder if we didn't start from 100% hydrogen. We ask:

- How did the stars form?
- How did the elements beyond hydrogen form?

Speculations as to the starting condition include:

- A uniform cloud of hydrogen gas molecules
- A neutron star
- A point source (singularity)

Properties of Hydrogen

Hydrogen can exist as a gas, a liquid, or a solid depending on its temperature and pressure. Science does not know how or in what form it was inserted into our Universe. By Ockham's Law, it is simpler to say all the hydrogen was a gas, bypassing the question of from whence came the energy to evaporate the hydrogen if it started from solid hydrogen or a pool of liquid. In the following discussion, we ask reasonable questions. Let us be alert to how many assumptions and unanswered questions we find. Let us also distinguish between what is a fact and what is an assertion.

Neutron Star

BDS proposed that the original entity was a neutron star that exploded in a Big Bang (BB) that resulted in a sea of intensely energetic neutrons half of which decayed to free protons and electrons within eleven minutes. Many of these protons managed to combine with some of the undecayed neutrons to form all the chemical elements by the stepwise addition of protons. What was left was hydrogen.

Comment: This is credible for the light atoms, but not for the 92 protons and the necessary number of neutrons to form uranium-238 and its isotopes in the short time available.

Question: Where did the neutron star come from? What held it together? No nucleus today has only neutrons. Is radioactivity therefore suppressed by very high pressures? Why did it explode?

Comment: Science has strong faith in the Doctrine of Random Events and a conviction that no matter how small the probability of an event, that event must have happened if Science is unable to conjure up another or accept a contrary view. However, it applies this Doctrine selectively.

Eventually, various size gas clouds of these elements formed. Gravitational contraction heated the clouds to 1E06 K (**TOE** says 5E06 K; **TDU** says 10E06) to permit hydrogen + deuterium to form helium-3. On exhausting the deuterium, contraction resumed, the temperature rose to permit conversion of the next heavier element, lithium, and so on. Finally the Bethe/Weizsäcker carbon-nitrogen-oxygen cycle our Sun uses started at a calculated temperature of about 20E06 K. The results are said to match the observed radiation rate from the Sun. The reaction can continue for a very long time because of the huge supply of hydrogen. **TDU** says this cycle occurs only in stars more massive than our Sun.

The **BDS** proposal was set aside because the process stops at atoms of mass 5, all of which have unmeasurably small half-lives. Let's see.

In principle, any particle can combine with any other, but the product may not be stable. It is easy to accept the proposal that there was so much hydrogen present, and that it would react with other elements from the BB to form stable nuclei. For example:

Hydrogen-1 + hydrogen-1 → hydrogen-2 (deuterium)
Hydrogen-1 + lithium-7 → 2 helium-4

But it could also react with the new elements.

Hydrogen-1 + hydrogen-2 → helium-3

And couldn't we also have:

Hydrogen-1 + neutron → hydrogen-2 (stable)
Hydrogen-2 + neutron → hydrogen-3 (decays)

Hydrogen-2 + helium-4 → lithium-6   (stable)
Helium-3   + helium-4 → beryllium-7 (decays)
Beryllium-7 + electron → lithium-7   (stable)

It was also postulated that a neutron star can be one of the end points of a star's life. This leads to a cyclic Universe: an original neutron star explodes and many stars are formed, many of which become neutron stars. Some claim to have detected them. Why don't they explode?

Comment: It would seem a further check of the literature on nuclear reactions is needed.

Uniform Cloud

<u>Initial conditions</u>

Once upon a time, there was a huge expanse of hydrogen molecules. We are not told how fast they were moving: was the energy in the cloud enormous or minimal? Or in what direction(s) they were moving. How did such a cloud of gas develop into stars and galaxies? We have three possibilities.

A. Infinite cloud of gas in an infinite Universe

It is suggested that gravitational attraction caused the mass to collect at least into an initial star.

Comment: This is implausible: there would be just as much mass on one side of a molecule of hydrogen as on the opposite side. There would be no net force pulling it in a particular direction.

Another suggestion: shock waves from novas and supernovas started the process.

Question: This is an illogical evasion of the question. What caused the <u>first</u> star to form and eventually explode? What is the medium transporting the shock waves through space? Or is the shock wave a coherent assembly of high velocity particles?

B. Finite cloud of gas in an infinite Universe

First, a finite <u>amount</u> of gas dispersed over an infinite Universe is meaningless, since effectively there are zero particles per unit volume.

Second, experience tells us that in billions of years any gas should have become uniformly distributed and infinitely dilute. Perhaps the force of gravity operated from the moment of emergence of the cloud of gas. It is then a real mystery how the entire cloud emerged from nothing, since the boundaries of the Universe, if any, were an infinite distance away.

> **Question:** Was matter and/or energy from another Universe <u>injected</u> into ours to the present average density? Accidentally, or on purpose? If we really want to be tough-minded about this, recall that energy can be converted into matter. So matter is a way to store energy. Perhaps in another Universe there is a process generating excess energy. Our Universe might be only a convenient garbage dump, like how we treat our oceans.

However, if such a cloud somehow did exist, its behavior would be similar to the cloud in Possibility C.

C. Finite cloud of gas in a finite Universe

There are two cases: the cloud was initially smaller than the Universe or the cloud filled the Universe. On Earth, gases in containers do not collect in a small portion of the container. On a cosmic scale, the gravity of an enormous quantity of gas might hold the mass in a portion of the Universe.

The Law of Gravity is thus offered as the god that causes order to appear in the secular Universe. Its action is opposite to the general tendency of things to run down, of potentials to dissipate, of randomness to increase, of every quantum level to be occupied.

> **Question:** Did the Law of Gravity come into existence before the cloud of gas, after it, or at the same time as the cloud? From whence came this Law? Did it create itself? If it is an intrinsic property of matter, what is in the structure of matter that can produce a Law reaching to infinity?

1. Initial conditions at the edge of the cloud

Suppose that the mass of our Universe at critical density was uniformly distributed in an initial sphere of radius R, the same as now: 14 billion LY, or 1.324E28 cm. By Newton's Law, the acceleration of gravity, g, at the surface of the sphere was:

$$g = (6.672-E08)(8.96E55) / R^2 = 3.408E-08 \text{ cm/s}^2$$

From the kinetic theory of gases (**KTG**), the root mean square (random) velocity of a molecule of hydrogen at 3K was 0.120 miles/second. Gravity will give the molecule a velocity component inward that will be 0.120 mi/s after travelling inwards a distance s:

$$s = v^2/2g = (19310 \text{ cm/s})^2 / 2g$$

$$= 5.47\text{E}15 \text{ cm} = 0.00578 \text{ LY}$$

and the time involved is $t^2 = 2$ s/g, from which

$$t = 5.67\text{E}11 \text{ s} = 18,000 \text{ years.}$$

> **Comment:** Despite any intrinsic kinetic energy of the molecules, the continuing force of gravity will still draw them inward at an increasing rate. But clearly we need a colossal sphere of gas to have this process work in any reasonable time.

The "fall" time varies as the radius of the sphere and the square root of s. If the radius was 1 million LY, the time to fall 1 LY would be 17 years.

2. Jeans criterion

In rejecting Creation, Science can explain the existence of our Sun and the stars only by assuming that (somehow) hydrogen accumulated by gravity. Jeans says that only very massive gas clouds can form stars by gravitational contraction, but does not explain how small stars like our Sun can form. (Somehow) a large mass breaks up into smaller ones, which then fall within his density criterion. No mechanism for this breakup is proposed. Our inquiry is not affected because we are considering a very large cloud.

Life history of protostar

A. Initial condensation process

As the hydrogen cloud condensed more or less in the center of the Universe under the unbalanced force of gravity, molecules collided more frequently. Energy was not lost since the collisions were perfectly elastic. However, motion means kinetic energy (KE), so the molecules had a temperature and they radiated energy as photons. Photons interact with matter, so the cloud became ever more opaque to radiation and the temperature rose even faster.

> **Question:** Even if the temperature of the cloud was initially uniform, it could not remain so. And surely some photons on the periphery of the

cloud would escape to the outer Universe. Since our basis is a large, but finite, Universe, how were escaping photons directed back to the cloud?

The molecules dissociated into atoms when the bonding forces were exceeded by the energy in the molecules. This is a gradual process since there is always a distribution of velocities, hence energies. **EST**, Hydrogen, says 96% is dissociated at 5000 K. We know there are hydrogen atoms at the Sun's surface at 6000 K.

In the next stage of heating, orbital electrons were stripped from the atoms to leave bare protons.

> **Question:** Electrons are supposed to be only waves around a nucleus. How does one strip a wave from an atom? How does a free wave become a free electron? During these changes, how does each electron still carry exactly the same quantity of charge?

Perhaps it is easier (better?) to think that the electrons have acquired too much energy to remain in stable configurations around the nuclei. They remain mixed in with the protons so the mass is still electrically neutral.

> **Comment:** It is indeed awesome to contemplate such a relentless force as gravity that can overcome even enormous increases in kinetic energy.

B. Deuterium formation

   1. Reaction description

The current hypothesis is that the elements are built up by thermonuclear reactions in the stars and not in the BB. Deuterium formation is now thought to be the first step, so we must understand it thoroughly.

$$_1^1H^+ + {}_1^1H^+ \rightarrow {}_2^2H^{++}$$

$$\rightarrow {}_1^2H^+ + {}_0^0e^+ + {}_0^0\nu^0$$

The subscripts are the numbers of nuclear protons and the superscripts are the masses (protons + neutrons). The initial product would have mass 2 and charge +2, but is unstable. It ejects a positron $({}_0^0e^+)$ to become a stable deuteron, the nucleus of deuterium. The positron mass, like that of an electron, is considered negligible in such equations. The neutrino mass is 0; it is there to conserve spin and carry off a bit of energy (somehow).

Comment: When two protons combine, somehow they know they must expel a unit of positive charge. Whatever makes up the mass of a positron was wholly contained in the mass of the two protons. Does half the mass and charge come from each? If it is a random event, the proton that has less mass must absorb enough energy to become a neutron.

The deuteron is "explained" as consisting of a proton and a neutron. The neutron is "explained" as a proton and an electron combined somehow. So two positively charged protons can on their own cough up half their total positive charge and create a negatively charged electron!

The positron produced is recycled: it quickly runs into one of the two electrons in the plasma from the reacting H atoms and vanishes as a photon of energy leaving the plasma electrically neutral.

2.   Reaction temperature

To overcome the large <u>repulsive</u> force between two positively charged protons, their kinetic energy and thus their temperature must be extremely high. **CVEP** says the proton-proton interaction can occur if a) they come within about one diameter of touching or b) the center-center distance is less than the proton radius. I read a) as a C-C distance of 2 diameters, or 5.6E–13 cm and b) as 1/2 diameter, or 1.4E–13 cm. **CVEP** also says the approach must be within 1E–13 cm for that is the range of the strong nuclear force.

Using the formulas relating kinetic energy, potential energy, and temperature, for 1E–13 cm, the temperature is 5.5 billion K. Other references give temperatures in the range 5 to 10 million K.

Question: Consider also that the plasma contains an equal number of electrons. Why don't <u>they</u> stick together? Well, the electron is the smallest unit of negative charge ever observed. It has never been seen to split off a smaller particle with a unit of negative charge to leave an entity that is electrically neutral with the mass of an electron.

Question: Why can a proton split off a positron? Surely this implies some kind of proton structure. Science talks of protons being made of quarks but doesn't tuck half a positron in each somewhere.

Question: In the ferociously energetic environment involved, is it not possible for two protons on a collision course to trap an electron between them? Well, it is a less probable three-body collision. But it should occur at a lower temperature because of the <u>attractive</u> forces of the electron.

3. Energy balance

Let's now look at the energy balance for the equation in Section B.1. Consider a high velocity proton A striking a stationary proton B. **NCE** gives the rest masses in atomic mass units (amu). We convert these to grams and then by $E = mc^2$ to ergs.

Table 14.1 Energy Balance for Deuterium Reaction

	Rest Mass		
	Amu	E–24 Grams	E–06 Ergs
Hydrogen	1.0078250	1.6735608	1504.1214
Electron	0.0005486	0.0009110	0.8188
Proton	1.0072764	1.6726498	1503.3026
x 2			3006.6053
Deuterium	2.0141022	3.3445513	3005.9328
Deuteron	2.0135536	3.3436404	3005.1141
Balance			1.4912

Description

• A hydrogen atom less its orbital electron yields a proton; a deuterium atom yields a deuteron.

• The KE of A at 10E06K adds 2.071E–09 ergs, or only 70 millionths of a per cent of the energy of the two protons. Yet it releases 720x as much energy.

• To achieve stability, the excited deuteron must eject a positron representing 0.027% of its energy.

Comment: Interesting. Somehow the deuteron cannot tolerate even the tiny KE of the proton and must create a positron whose mass is 1/1836th of the proton that can capture 55% of the energy released and the entire charge of one proton.

4. Particle collisions

Consider three kinds of particle interactions:

a) An electron falls from infinity to orbit around a proton;

b) A high velocity proton falls from infinity to collide with and join another proton; and

c) A high velocity proton falls from infinity to the point closest to another proton.

We make the following observations.

• Earlier, we talked about a). But the texts made no mention of the Conservation of Momentum then. Since the electron initially was at rest at infinity, it had neither kinetic energy nor momentum. The appearance of KE was explained, but the angular momentum in orbit was not. What was its origin?

CVEP says the orbit entered depends on the direction of the velocity relative to the direction of the electrostatic attraction. It omits comment on the case of zero angle between the two directions.

• Similarly, for b) we need an explanation of what happened to the initial linear momentum of the high velocity proton.

• The interaction c) is a transient one. If its energy is insufficient for fusion, the approaching proton is slowed to zero velocity and then is repulsed back to infinity. However, the same question arises: as its velocity changes, where does the momentum go to or return from?

Ponder on the condensation process to achieve the dense packing and create the extremely high temperatures. It is the collective influence of the core atoms reaching out to pull more atoms toward the core. Like squeezing oneself to death. The origin and nature of this force is nowhere explained.

5. Spin conservation

In the equation as written (**TDU**, p A15), mass, charge, and lepton number are conserved. If the deuteron is +2, baryon number is conserved . The spin of the two protons adds up to +1, 0, or −1, so the spin of the products must also add up to these figures. The deuteron spin is 1 (**EST**, Spin). Presumably this means ±1.

The spin of electrons is ±½, so that of the positron must also be ±½; the neutrino is ±½. So there seem to be six varieties of the above reaction depending on how the spins are matched up. E.g., if the protons are both +½, the products must add up to +1. Two cases are if the deuteron is +1, while the positron and neutrino have opposite spins.

The reaction is an attempt by the star to readjust its distribution of matter and energy to accommodate the higher pressures, kinetic energies and frequency of collisions. As its core grows, something forces it to disgorge neutrinos which, being neutral electrically, readily escape from the star. (Somehow) they carry away lots of spin, whatever that is.

> **Question:** The angular momentum of the hydrogen atom is said to consist of the angular orbital momentum of the electron plus the spins of the proton and the electron on their "axes". When a hydrogen atom is ionized, what happens to the <u>orbital</u> angular momentum of the electron? Or is this another reason why the orbital concept just doesn't hold up? Or perhaps the particle spin assumption is wrong. . . .

> **Question:** How does a photon differ from a neutrino?

> • Neither has charge nor mass.

> • The velocity of a photon is c, so its relativistic mass can have a finite value. This might apply to the neutrino, and some speculate that the neutrino has a tiny mass.

> • Both are said to have intrinsic angular momentum but the spin assigned to the neutrino is ±½ while that of the photon is 1 (±1?).

> • There is an antineutrino, but the photon is said to be its own antiphoton.

> **Question:** Science shows it is as meticulous as Religion in trying to tie down loose ends in the quest for consistency. Why was the antiphoton invented? To preserve the illusion of symmetry?

So a particle is a wave except when it is not and it may have mass or not. Science moves bravely on from these ambiguities, declaring boldly that it is consistent and affirming in full faith that this is a solid basis on which to model the Universe.

C. Helium formation

The deuterium formed goes on to produce helium. A deuteron plus a proton forms stable helium-3. Two of these combine to helium-6 which splits off two protons to give stable helium-4. The process is stable (i.e., it does not result in a runaway explosion) since the contraction due to gravity is just balanced by the expansion due to heating. Overall:

$$4 \, p^+ \rightarrow {}^4He^{2+} + 2 \, e^+ + 2 \, v^\circ + 2\gamma^\circ$$

We then do a rest mass balance. For the alpha particle, we use the mass of a helium <u>atom</u>, since we assume a positron has the same mass as an electron.

Input:	$4 \, p^+$	6.6906E–24 g
Output:	$1 \, {}^4He^\circ$	<u>6.6466E–24</u>
Mass loss:		0.0440E–24 g

The loss is 0.66% of the mass of the four protons. By $E = mc^2$, the energy released is 3.955E–05 ergs per helium nucleus formed. This is equivalent to 24.7 MeV.

For every four protons reacting, two neutrinos appear. Our Sun releases 3.85E33 ergs/s so the above reaction occurs 98E36 times every second, releasing neutrinos at twice this rate. By a calculation similar to that for radiant energy in Ch. 13, every 1 cm^2 of the surface of the Earth is bombarded by 70 billion neutrinos per second. We need some discussion of their impact on humans, and what happens to them and their KE. Do they confer spin on something?

What happens to the helium made? A star is gaseous, so its helium should be distributed uniformly. Yet our Sun shows an outer atmosphere with only about 7% helium. By current theory, the deuterium reaction should produce a Universe with 25% helium. Therefore, there must be relatively more helium in the interior of our Sun as alpha particles.

D. Red giants

Some stars have low temperatures but large luminosities, which suggests they have large surface areas. Their light is redder, so they are called red giants. We will continue with the history as if the protostar were on the HRD.

As the star develops a helium core, its rate of heating and its temperature decrease. Gravitational contraction resumes, causing the outer layers of gas to heat up and expand. The diameter of the star expands by as much as 100 times. The star leaves the main sequence as a red giant.

Science says that when nuclear reactions resume in the core, it is by the triple alpha reaction: two helium-4 atoms form the extremely unstable beryllium-8. Since carbon-12 can be decomposed to three helium-4 atoms, the reverse reaction <u>must</u> be possible, by the Creed of Random Events. Be-8 must last long enough to join a third helium atom to form some carbon-12.

Question: The logic is: the only hypothesis accounting for the existence of carbon-12 is the triple alpha particle reaction. So it must have happened. Does the thermodynamics of the reaction result in a frequency of its occurrence sufficient to satisfy the observed amount of carbon-12?

Comment: However, it requires highly improbable events: a triple collision, or a double collision plus an essentially simultaneous followup collision. A more credible path seems possible.

One and one is two.	This makes deuterium.
Two and one is three.	This makes helium-3.
Three and three is six.	This makes helium-6.
Six and six is twelve.	This makes beryllium-12.

Be-12 decays (0.012 s) through boron-12 to carbon-12 (0.020 s). Helium-6 has a half-life of 0.8 s. In each, an electron is ejected from the nucleus. He-3 and C-12 are stable. The series seems more probable since it involves only two-particle collisions and radioactive decay of individual nuclei. Why it is not offered as the path to carbon-12?

A loose end is how are lithium, beryllium and boron formed. Answer: by unspecified secondary processes. **SSE** in a brief passage concludes that the origin of lithium is still uncertain, but is thought to have been made in the Big Bang. Anyway, it is consumed rapidly above 2 million K, making helium.

E.   Formation of heavier elements

Some process other than just adding protons via hydrogen is needed to explain how even heavier elements are formed. Observation of existing stable nuclei permits the inference that neutrons must be added. Or, some electrons must be forced into close association with the nuclear protons.

It is postulated that heavy elements are built up by adding alpha particles. Atom building can then go on to iron (mass 56), beyond which energy is absorbed and so cannot generate the heat needed to stabilize the mass. It is postulated that supernovae are required to build elements to uranium and beyond.

Before we move on, we can make an observation about nuclear structure. The building blocks seem to be protons (P), electrons (E), and neutrons (N).

Charles H. Peterson

The latter has slightly more mass than the one proton plus one electron. Let's ignore this slight difference for the purpose of our question and treat it as one proton plus one electron. So we count the total positive charge (protons) and the total negative charge (electrons). The baryons in the first two elements are in Table 14.2 for illustration.

There are no simple rules for nuclear stability.

- A proton can exist by itself (hydrogen-1); a neutron cannot.

- Two protons or two neutrons cannot bond.

- A proton/electron ratio of 2 is stable (H-2, He-4, Li-6, B-10); Be-8 (8p + 4e) is very unstable.

Table 14.2 Structure of Isotope Nuclei

Element	Baryons		Count		Ratio	Stability*		
	P	N	P	E		S	D	N
Neutron	0	1	1	1	1.00		X	
Hydrogen	1	0	1	0	∞	X		
	1	1	2	1	2.00	X		
	1	2	3	2	1.50		X	
	1	3	4	3	1.33			X
	1	4	5	4	1.25			X
Helium	2	0	2	0	∞	X		
	2	1	3	1	3.00	X		
	2	2	4	2	2.00	X		
	2	3	5	3	1.67			X
	2	4	6	4	1.50	X		
	2	5	7	5	1.40			X
	2	6	8	6	1.33		X	
	2	7	9	7	1.28			X

*S = Stable; D = Decays; N = Nonexistent

- A p/e ratio of 3 is stable (He-3), but Be-6 and C-9 are unstable; B-8 (8p + 3e) is also unstable.

- Helium-5 (5p + 3e) breaks down immediately.

- A p/e ratio of 1.5 is unstable: H-3, He-6, Li-9, Be-12, and B-15.

- A p/e ratio of 1.33 is unstable: H-3, He-8, Li-12.

These observations suggest that there is either a repulsive force due to the electron or there are nuclear structures into which excess electrons cannot fit. Perhaps someone may be inspired to think up a model accounting for these differences in nuclear stability.

F. Events after exhaustion of hydrogen fuel

1. Gravitational contraction

Eventually, the hydrogen fuel is gone and the gravitational contraction resumes. The next change depends on how much mass remains. Suddenly and unpredictably the luminosity of a few will show an enormous increase that then gradually diminishes. At least 20 stars of our galaxy explode every year as novae and supernovae. Why? **BDS** says we don't know.

2. Chandrasekhar limit

Calculations by Chandrasekhar and Kothari showed there is a critical value for the residual mass, which is the mass after nuclear reactions cease. Various sources differ on the final characteristics.

- If the residual mass is less than 1.4 solar masses, the star will become a white dwarf with a radius less than a few thousand miles. The range cited for density is 5E05 to 1E09 g/cm^3. **BDS** says 2E05 g/cm^3. The surface temperature is 4000 to over 100,000 K (**EST**, White Dwarf Star).

- If it is less than 0.7, the collapsing star becomes a neutron star with a radius of ten miles and a density of a million times greater than that of a white dwarf.

- If it is larger than 1.4 times our Sun, there will be unlimited (?) contraction to a nuclear state of matter.

• If the mass is greater than 3 solar masses, we get a black hole. These are smaller in volume than neutron stars, but denser. However, **TDU** says it will lose all mass greater than 3 Solar masses in a supernova explosion.

Black holes are less clearly defined. In **TDU** we read of the Schwarzschild radius, R, at which the gravitational field of a collapsing star becomes strong enough to prevent photons from escaping. I.e., the star has just become a black hole. For a mass M it is:

$$R = 2\ GM/c^2$$
$$R/M = 2\ G/c^2 = 1.485\text{E}{-}28 \text{ cm/g.}$$

If the star has a mass of n times that of our Sun:

$$R = (1.485\text{E}{-}28 \text{ cm/g})(1.989\text{E}33 \text{ g})(n) = 2.954\text{E}05 \text{ n cm.}$$

For a spherical black hole, the mass density is:

$$\rho = M\ /\ [(4\pi/3)(R^3)] = [3/(4\ \pi)](M/R)\ /\ R^2 = 1.842\text{E}16/n^2 \text{ g/cm}^3,$$

For n = 1, this is about 125 times the density of a proton. The radius of our Sun would have to decrease by a factor of about 240,000. But the factor n indicates a strange result. If a galaxy having 100 billion stars like our Sun collapsed, when its density reached $1.84\text{E}{-}06$ g/cm^3 it would be an incipient black hole. Thinking of black holes as very dense regions is somewhat understandable, but the concept seems to lose meaning if it also includes very low densities.

3. Crushed matter

BDS postulated a crushed state of matter. At normal pressures, the electronic shells of the atoms are impenetrable. As pressure rises, the atoms would be forced closer together until they touched. On further compression, their electron shells might interpenetrate. Bare nuclei and free electrons would be rushing around. Perhaps this is "electron degeneracy".

Question: How could they rush around if there is no space to rush around in?

• Electrons from atom B intruding into atom A would be closer to the nucleus of the invaded atom, so B electrons would be held more tightly by the electrostatic attraction of the A nucleus, and vice versa.

- How is electron degeneracy pressure reconciled with the notion that for a mass of contracting gas more than 1.4 times that of our Sun, the contraction results in a neutron star?

BDS cites Kothari's estimate of 150 million psi as the pressure level of atom crushing. The white dwarf companion of Sirius is an example; however, it still has a 10,000 K surface temperature.

Comment: In Ch. 11 we estimated the strong nuclear force as 690 trillion trillion psi. This is in conflict with the crushing pressure.

4. Electron "gas"

Since the electronic shells were no longer organized entities, further compression is possible. The electrons (somehow) exert a force opposing that of gravity and so are not crushed. (Don't the protons have a similar objection to being crushed?) It is asserted that as a degenerate gas, the pressure of the plasma no longer depends on the temperature.

Question: Is there any experimental proof of this?

Fermi's work showed that the pressure of this electronic "gas" would increase inversely as the 5/3 power of the volume. I.e., if the volume decreased by a factor of 2, the pressure would go up by a factor of 3.17. His argument then supposed the mass of each volume element to be doubled. Gravitational forces between volume elements would be quadrupled, overwhelming the "gas" pressure. He didn't say through what mechanism such doubling would occur.

Question: Where are the protons in all this? And we have already learned that the electrostatic attraction or repulsion is many orders of magnitude greater than the gravitational attraction.

Landau said that the contraction must stop when the separated nuclei and electrons touch. The electrons will stick to the nuclei to form a continuous nuclear substance that will be electrically neutral. The electrons have been forced into at least a close association with the protons to form neutrons. The result cannot be predicted without some concept of proton structure. Consider what Science asks us to believe.

- In a positron/electron collision, two gamma rays will appear in place of their mass.

295

- If electrostatic attraction pulls an electron into an orbit around a proton, we have a hydrogen atom.

- If the proton becomes closely associated with the electron, we have a neutron, not annihilation.

- An electron can orbit a positron to form a positronium atom. Presumably, the reverse is possible.

Question: What governs the choice of final states? From whence came these selection rules?

5. Gravitational energy

F. Zwicky postulated that this nuclear mass would eventually collapse and release an enormous amount of gravitational energy.

Question: What does "collapse" mean? Why would it collapse? Is the gravitational energy the PE of the individual particles?

The internal radiation pressure would expel the outer portions of the star as an expanding shell. Sort of a cosmic regurgitation. It isn't clear what becomes of the core. **FG** says it becomes a pulsar. **BHT** says it would likely go to a neutron star or a black hole. A variety of elements form by recombination of particles into neutrons and atoms, and are flung out into space.

G. Second generation stars

Since all the mass in the Universe was in a single star, it clearly was much more massive than our present Sun (5E46 x). By the above criteria, it became a supernova (hypernova?). We have many more questions.

- Why wasn't all the expelled material dragged back to the extremely dense core remaining? Did the core serve as a nucleus for a new star? Is this the Great Attractor?

- How did the matter flung out into space condense into new stars? Was the primordial star completely symmetrical or were there some aggregations that meant nonuniformities in the final distribution?

- Did the total mass of all the elements get formed in this first explosion, or did the Universe have two or more generations of stars?

- Star life seems to be shorter as mass increases. Perhaps the lifetime of the primordial star was of the order of only millions of years. The second generation of stars, working with much smaller masses, should have taken longer to form.

- It is asserted that new stars are continually being born. Since the process seems to take millions of years, what has Science observed in 100 years that supports this assertion?

Overall, though, the above scenario for Possibility C does not seem impossible.

Singularity

In this section we will examine the Book of the Redshift and the Doctrine of the Expanding Universe, which seem to have gathered almost universal support.

A. Basis for Expanding Universe

The sole basis of this explanation for the origin of the Universe appears to be the redshift observations. Except for nearby objects, all interstellar bodies appear to be moving away from us. Scientists infer that in the past they were all closer and assume they can extrapolate this trend billions of years into the past until all matter and energy was concentrated in a single point of infinite density.

Question: Large extrapolations become increasingly questionable. How far back can one go in time? The farther the more likely some other forces or events were involved.

But there is a puzzle. If our Universe is 14 billion years old, if all the stars are moving away from each other, why do the galaxies still have disc and spiral formations? How can we get a redshift for an entire galaxy if all the stars are moving away from each other?

B. Assumptions

The Big Bang hypothesis has at least nine assumptions.

- In the beginning there was an original enormous amount of something (matter, energy, or?).

- The something was concentrated into a single dimensionless mathematical point.

- Space was also compressed within this point.

- The point was (a) contained in an infinite something that is not space or (b) contained in nothing, which is incomprehensible and probably ridiculous.

- The something was either of unknown instantaneous origin or it existed for an indefinitely long time.

- At some unknown stage of its unknown existence, the point exploded for some unknown reason.

- Its subsequent history was an expansion of space carrying photons with it.

- Matter did not form until some later, but still relatively early, time.

- Matter was carried along with the expansion.

Comment: Absolutely no proof exists for any of these assumptions to distinguish them from fiction. The image of a nothing (space) carrying something (photons, particles) with it reminds me of the scam in Andersen's fairy tale of "The Emperor's Clothes".

Comment: Various writers also speak confidently of times as short as 1E–43 seconds. I have no wish to become an expert on these concepts; I merely wish to understand the <u>basis</u> for what is being said.

Eventually I found in **TEU**, Appendix A, information that might explain the origin of these incomprehensibly minute times and dimensions. It seems that some of the fundamental constants of our Universe can be combined in various ways.

$$\text{Planck time} = t_{Pl} = (hG/2\pi c^5)^{0.5} = 5.39\text{E}{-}44, \text{ or } \sim 1\text{E}{-}43 \text{ s}$$

$$\text{Planck length} = l_{Pl} = (hG/2\pi c^3)^{0.5} = 1.616\text{E}{-}33, \text{ or } \sim 1\text{E}{-}33 \text{ cm}$$

Why should these groups have any particular significance? As we noted earlier that curiously the Earth's acceleration of gravity in astronomical units was about 1 LY/yr^2, so also are these groupings only curiosities.

Comment: Can Science even measure such minute quantities? The atomic clock is said to be good to 32 parts per quadrillion. In contrast, the Planck time itself is 54 quadrillionths of a quadrillionth of a quadrillionth of a second, while the Planck length is 1.6 quadrillionths of a quintillionth of a centimeter.

C. Causes

Science undertakes absolutely <u>no</u> discussion of the cause of the Big Bang. It wiggles away from the question by saying time began with the BB. Maybe <u>our</u> time began. This does not mean there was no prior time. A cause is implicit in what Science argues:

- There was a Cause of unknown nature and origin <u>before</u> our Universe existed;

- At some point, this Cause was the cause of the appearance of our Universe, i.e. our Universe had a beginning; and

- The Universe operates according to mechanical "laws". Their origin and time of appearance are unknown but they seem fixed.

Religion agrees:

- There was, and is, a God of unknown origin;

- At some stage of His existence, this God elected to create our Universe; and

- Its subsequent history was in accord with laws that He set.

An important difference is that Science is <u>helpless</u> before the question of ultimate origins and its only course is to refuse to speculate on them whereas Religion forthrightly follows the evidence logically to the inference that there was an intelligent but inscrutable Creator.

D. Point source

In view of the ideas on a crushed state of matter and electron degeneracy and perhaps proton/neutron degeneracy, and in view of the number of protons in the Universe, a point source is grossly inconsistent.

Illustration: If the 8.96E55 g of total matter (by critical density, Ch. 13) is present as protons at a density of 1.455E14 g/cm^3, the total volume of the protonic material is 6.16E41 cm^3. Using a packing factor of 0.74 for spheres, meaning that 26% of the volume is empty space between the protons, they could fit in a sphere 1.167 billion km in diameter. Very much less than a light-year (0.0001234 LY), but hardly a point source.

E. Time sequence

Let's follow the time sequence offered.

    1. Initial composition

The BB is supposed to have immediately filled the Universe, whatever size it was initially, with a plasma of various elementary exotic particles (**EST**). **AJASC** says only quarks were present.

> **Question:** Particles? No waves? The "temperature" was supposedly unimaginably high. Perhaps only photons were around, i.e., the starting point was intensely concentrated energy.

> **Question:** The temperature of what? Temperature is the kinetic energy of <u>matter</u>, not photons.

> **Comment:** "Exotic" is not defined, and the dictionary does not help because it offers "foreign" or "unusual". "Exotic" seems to refer to the mesons and hyperons that have lives shorter than a will-o'- the-wisp. The lives are no longer than nanoseconds. How do they qualify as elementary building blocks?

    2. Expansion rate

EST, Big Bang theory, asserts the Universe expanded by a factor of billions within a very short time.

> **Comment:** There is a serious problem with dimensions.

> • If the initial condition was a true singularity, its dimensions were zero. A billion times zero is still zero. Do we suspend the laws of mathematics?

> • The factor of "billions" could be as much as 1000 billion. If the singularity was a sphere one mm in diameter, a 1000 billion times that is still only a million km, or 1 ten-millionth of a LY.

AJASC asserts that the Universe doubled in size every 1E–35 seconds.

Questions: Well, we certainly can pick an expansion factor arbitrarily, but how do we know:

- that it applies every 1E–35 seconds (why not every 1E–43 seconds?);

- that it continues for as long as Science arbitrarily decides it must do so; and

- that the doubling stops after 1000 doublings? Why 1000? Why did it stop at all?

Comment: The expansion factor is 2^{1000}, or 1.07E301, and a 1-mm "point source" expands to 1.07E301 mm. Now only 93 doublings would reach 1.05E09 LY. This is nearly the distance to the constellation Hydra. Every 10 doublings thereafter multiplies the distance by 1024. Has anyone proposed the Universe is even a trillion LY in radius?

3. Cooling

Science asserts that expansion of the Universe caused cooling. **BHT** states that for a doubling of volume there is a halving of temperature. The basis for this is not given, leaving us to wonder about it.

    a.   The Second Law

To cool an entire Universe from 10 billion to 1 billion K in five seconds **(EST)** requires an incomprehensibly enormous rate of heat transfer. How? We don't know if our Universe is an isolated system; Science says it is. So where did the heat go? By the Book of Thermodynamics, the Second Commandment for an isolated system is Conservation of Energy: heat (energy) is conserved if a photon moves from State 1 to State 2:

$$E_2 - E_1 = 0,$$

but if the energy changes, Planck's Law requires a frequency change:

$$E_2 - E_1 = h\,(v_2 - v_1)$$

$$= (6.626E{-}27 \text{ erg-s})(428E12 - 750E12)\text{Hz}$$

$$= -\,2.134E{-}12 \text{ ergs/photon}.$$

The frequencies are from Table 9.1 and correspond to stretching a blue wave bordering on the ultraviolet to a red wave bordering on the infrared. This is a loss of 48% of the energy. Is it gobbled up by the space monster as it strives to expand? One answer is that there are <u>more</u> red photons than blue. So is our Universe gradually filling up with low energy photons?

AEU asserts that the expansion of the Universe caused another process. Whenever photons and matter collided, both lost energy that went nowhere. In the Universe as a whole, energy is not conserved!

> **Comment:** Notice how a question is answered by an assertion. How can we build a concept of our Universe if there are always exceptions to the Commandments? Who is the prophet who will tell us of the exceptions and to what do they apply? I would rather Science retain the idea that the energy goes somewhere, but admit that it doesn't know where.

b. Gas expansion

Perhaps the particles are being treated like molecules of a gas expanding freely into a vacuum. Unfortunately, we do not have data for very high pressures and temperatures. Nor do we have data for protons and electrons, let alone exotic particles.

> **Question:** Could the drop in "temperature" result merely from fewer molecules passing through a given area perpendicular to their path in space?

c. Red shift

AEU says the inverse relationship follows from the interpretation that the red shift means the Universe is expanding. **AEU** asserts that light travelling through an expanding Universe is expanding with it, so it is appropriate for the wavelength to vary directly with the Universe radius. In quantum theory, a wavelength is closely associated with temperature through the quantity λT. Wien postulated that if λ was the wavelength at the maximum emission intensity, this product was a constant. It is then a simple matter to infer that temperature varies inversely with radius.

> **Comment:** T in the Planck equation is the temperature of the <u>source</u>, not that of the photons. Here Science treats photons as particles and as waves.

d. Entropy

Eventually, I found another explanation. **TEU**, pp 65-68, shows that, assuming the Universe is homogeneous and isotropic, and that thermal equilibrium persists as the Universe expands, S, the entropy per unit comoving volume, is conserved. S is shown to equal aT^3R^3, where R is the radius of the Universe and T is its uniform temperature. If a is constant, T must vary as $1/R$. The derivation is impossible for a layman to follow: terms are not clearly defined nor are units given. There is a consequence: it in effect postulates a new concept of temperature based on an inverse relationship to distance. Out with the old, including the concept of absolute zero based on kinetic energy.

4. Elementary particle formation

Within 1 microsecond, all exotic particles and quarks were incorporated into various fundamental particles (**EST**, Big Bang theory).

• **AJASC** agrees on the timing, and adds that by 1 s the neutrinos no longer interacted with other particles. In what way did they interact before 1 s?

• **BHT** says this stage was reached in 1 second, and the temperature was 10 billion K. The particles are affected only by the weak force and gravity.

Question: How were these estimates made?

• The statement is glib. The fundamental particles are the electron and possibly the electron neutrino. The proton is made of quarks. The neutron is made of an electron, a proton, and an antineutrino. No exotic particles exist in any of these.

• When exotic particles decay, they yield particles of smaller masses, neutrinos, gamma rays, and electrons and positrons. No stable exotic particles.

5. Formation of atomic nuclei

At 5 seconds, the temperature was only a billion K and the Universe contained electrons, positrons, neutrinos, antineutrinos, and photons but relatively few protons and neutrons. **BHT** says it took 100 seconds. The protons and neutrons could no longer escape the strong nuclear force, and nuclei formed.

Question: What is the nature of the strong nuclear force and when was its origin?

303

Comment: We are told the range of this force is no more than 500 quadrillionths of a centimeter, which is less than twice the diameter of a proton (See Table 11.3). But how far apart are the particles?

TFE, p 326, shows a figure in which the recombination time is at about age 1 million years and the Universe radius is about 1E24 cm (1.06 million LY). The 5.36E79 particles in the Universe (at critical density) each occupied a volume of 78E–09 cm^3. As a sphere, its radius is 2.65E–03 cm. The proton spacing is thus about 6 billion times the range of the strong force. How <u>did</u> the protons get together? All the particles are moving away from each other.

It is then asserted that nuclei would form as discussed under Neutron Star. About 25% of the protons and neutrons would be combined into helium.

6. Recombination

At about 1E06 years, the temperature was 3000 K, and <u>suddenly</u> (my emphasis) protons and electrons combined to form hydrogen atoms. **AJASC** says 100,000 years.

Question: Why 3000 K? The binding energy of the orbital electron in the hydrogen atom is – 13.6 eV (Exhibit 11.1, Equation 10). So the ionization temperature is:

$$T = (2/3)(KE)/k$$
$$= (2/3)(13.6 \text{ eV})(1.602E{-}12 \text{ ergs/eV}) / (1.381E{-}16 \text{ ergs/K})$$
$$= 105{,}200 \text{ K}$$

At this temperature the energy in the atom exceeds the binding energy and the orbital electron parts company with the nucleus. However, 3000 K is near the temperature at which hydrogen molecules dissociate into atoms.

F. Location of the original explosion

1. Point source?

Our Universe is vast by human standards. Do we have any way of finding the site of the original explosion?

Let us use as a model a point source for the Universe. Let us also assume the explosion propagated spherically and uniformly. This means that every stellar object has moved and is moving radially away from the center of the explosion. Let's call this point O for origin. So the distance <u>between</u> objects is continually increasing at least because of the divergence of the radial paths. Consider three statements made by Science:

a) The Universe was originally concentrated in a single point.

b) From the corrected Sandage data, the Hydra cluster is 3960 million LY from the Earth and receding at 60,900 km/s, or 0.20 LY per year.

c) The velocities have remained constant for at least 3.96 billion years.

These three statements are inconsistent.

- First, the Relativity correction at 0.20 LY/yr is only 2%, and 0.5% at 0.10 LY/yr, so such effects can be ignored.

- Next, the recession velocity of Hydra is not its total velocity. That we cannot determine. If Hydra is fleeing from point O, so is the Earth, and we have measured only the <u>difference</u> in the components of velocities <u>in the line of sight</u>:

$$V_H - V_E = 0.20 \text{ LY/yr}$$

This is easily shown in two dimensions. Draw a large L and label the corner O for origin. Mark a point E at an arbitrary distance up the vertical leg from O. OE represents the distance the Earth has travelled since the BB. Mark a point H anywhere in the lower right to represent Hydra. OH is the direction and distance Hydra has travelled in the same time.

Draw the line EH. This is our line of sight. Actually it isn't, because what we see is the position Hydra had when its light rays started toward us. But ignore this detail for now. Extend line EH a distance proportional to the observed (line of sight) velocity of Hydra ending in point F. From F draw a perpendicular to EF to intersect OH, extended, at G. FG is the component of Hydra's total velocity <u>across</u> our line of sight, HG being proportional to the total velocity. We ignore the third component perpendicular to the plane of the paper. We make some interesting observations.

- As drawn, Hydra will have a large component across our line of sight, which clearly shows Hydra is not fleeing from us, but from some other point.

Question: How did Hydra get so far ahead of us if we started from the same point?

To say the stars are all moving away from us is to continue the medieval anthropocentric notion that <u>we</u> are at the center of the Universe and <u>we</u> are stationary.

> **Comment:** Recent work on the background radiation in the Universe suggests it might be used as a reference for <u>absolute</u> velocities (Einstein, are you listening?). Our local group of galaxies is moving at about 600 km/s relative to it (**TDU; TFE**).

TFE says the photon background is slightly blue shifted in the direction of motion and red shifted in the opposite direction. This is interpreted as a velocity relative to the microwave background and is attributed to a gravitational attraction from some unknown source.

> **Question:** Why? Is there some way of detecting the <u>increase</u> in this velocity this force would produce? It seems simpler to assume that any velocity is that which we had in the beginning. There is a myriad of gravitational attractions from every structure in our Universe.

2. Distance travelled

The velocity of Hydra can be anything up to c, the velocity of light, which is 1.00 LY/yr. For that matter, the Earth may be receding from point O faster than Hydra. Why don't we consider ourselves to be at the edge of the Universe?

Both bodies are supposed to have started from the same point. How far have they travelled in 14 billion years? We don't know. Let's assume that the two are moving in exactly opposite directions. Science does not yet want to concede that the Earth was the site of the original explosion, i.e., Creation.

Let's assume that each is travelling at the same velocity, 0.10 LY/yr, which preserves the observed separation velocity. In 14E09 years, each will have moved 1.4E09 LY from point O. Their separation is then 2.8 billion LY. Not 3.960 billion LY.

To make the separation equal to the observed value, point O has to be expanded so that both Hydra and the Earth were on the surface of a sphere 1.16E09 LY in diameter. This is the size of the sphere in which all the stuff that eventually made up our Universe was contained. It is <u>not</u> a mathematical point.

Of course, the distance to Hydra observed today is that existing 3.960 billion years ago. Both bodies have moved unknown distances, but the separation has a component in the line of sight that has increased by 792E06 LY since the first light started toward us 14E09 years ago. So it seems we must think in terms of a "point source" 1.16 billion LY in diameter.

This does not work with the other Sandage data. The distance the other constellations have travelled in 14 billion years is figured from their recession velocities. We find different diameters for the "point source". But this does support the idea that the nearer stellar bodies have been slowed down by gravity.

3. Deceleration of Hydra

What is the acceleration of gravity from the "point source"? Let's assume its mass M was 8.965E55 grams (Ch. 13) corresponding to the critical density and its radius was R_0 cm. Its acceleration of gravity g was:

$$g = GM/R_0^2 \text{ , cm/s}^2$$
$$= 6.683\text{E}12/R_0^2, R_0 \text{ in LY}$$

Even though Hydra must have a great deal of mass, we assume it is negligible compared to the mass of the rest of the Universe. Let's use 580E06 LY, as in §F.2., as the radius of the original "point source".

Since the Big Bang was supposed to have been an incomprehensibly enormous explosion, it is reasonable to suppose the fragments produced left the site at the speed of light, c. From Equation 8 in Exhibit 14.1:

$$A = c^2 / 2GM = 7.512\text{E--}29/\text{cm} = 7.108\text{E--}11/\text{LY}$$

$$(R_0)_{max} = 1/A = 14.07 \text{ billion light-years.}$$

Mathematically, this is the maximum value of R_0 for the values of M and v_0 chosen. If R_0 were greater, g would not be large enough to stop Hydra. Curiously, an average figure for the radius of our Universe is the same: about 14 billion LY. This also happens to be the Schwarzschild radius below which the mass M becomes a black hole (Protostar, § F.2). We also find:

$$s_{max} = A(580\text{E}06)^2 / [1 - A(580\text{E}06)] = 24.94 \text{ million LY}$$

$$t = v_0R_0(R_0 + s_{max})/GM = 1.574\text{E}15 \text{ s, or } 49.9 \text{ million years.}$$

So the g force would have slowed Hydra to a stop in 50 million years in only 25 million LY. The exploding star would not have dispersed much. This phenomenal result is due to the persistence of g. These numbers suggest that the gravitational attraction of 5.36E79 baryons should have stellar bodies receding much more slowly than they are. And this is even more true of the closer galaxies.

> **Comment:** The unceasing power of gravity suggests that, for the Universe to have become as large as it is, the force of gravity could not have operated over the whole time. This inference is also consistent with the hypothesis that in the beginning there was nothing but energy.

Let's state it explicitly: for some period of time after the Big Bang, if there ever was a BB, there was no force of gravity. Photons raced out to vast distances until matter began to form, and then the various forms of matter began to slow down.

> **Question:** The energy of the photon reflects the temperature of its source and photons don't interact. So again, how did the photons ever "cool off" to form matter?

We might have expected that Hydra would be so far from the site of the original explosion and the rest of the matter in the Universe that its velocity would be essentially unaffected by that matter.

> **Question:** Why do distant bodies appear to be moving faster? Perhaps we should reverse our interpretation: they are moving faster because they are closer to an attractive force coming from all directions.

4. Sphere of packed protons

Let's suppose the contraction under gravity can continue. In § D, we calculated the diameter of a sphere of densely packed protons for the mass of the Universe. Its gravitational force on a surface proton would be 29 nanonewtons. However, Coulomb's Law shows the repulsive electrostatic force is 36E27 newtons. Gravity cannot overcome the electrostatic force. When and how does the strong nuclear force postulated to hold baryons together in atomic nuclei come into play?

We can look one step further. In Exhibit 14.1, we find by Equation 13 there is a minimum radius to satisfy merely the condition of dense packing. But at this radius, the force of gravity is 230 trillion times the one just calculated. Can gravity (and the strong force) compress protons beyond dense packing?

> **Comment:** Perhaps the "strong nuclear force" is nothing more mysterious than the electrostatic attraction between proton and electron, even though the latter is somehow tied up in a neutron.

G. Microwave radiation

Much is made of the 3K isotropic microwave radiation found in our Universe. A. Penzias and R. Wilson in 1965 happened to pick up radiation well beyond the infrared at a 7-cm wavelength on their antenna. It was the same strength at any time during the day, in any season, and in any direction.

Further observations revealed radiation at other wavelengths matching the emission curve for a perfect black body at 2.75 K. This was interpreted as the remnant of BB emissions. Confirmation took a year of computer analysis to eliminate all other radiation received by the COBE satellite.

Question: Were these were singled out because they happened to fit the hypothesis of Big Bang radiation? The reports should have also described all the other radiation and their significance. What were the "temperatures" of these?

Besides filtering out "extraneous" radiation, the data had to be corrected for the Earth's motion. It would be of interest to know how large the corrections were compared to the final data. It is a tribute to the technology that this was at all possible. It is again of interest how such important conclusions rest on such minute measurements.

H. Isotropic nature of radiation

Even Science wonders why the microwave radiation is isotropic, i.e., looks the same in every direction. The standard assertion seems to be that, once the Universe became transparent to radiation, the radiation went in all directions, while (somehow) cooling off.

Objection: Radiation is said to travel in straight lines at the velocity of light. If the Universe is 14E09 years old, the radiation should long since have receded into the distance. That is because our velocity is much lower and everything started from the same point. If we look in any direction, we should see nothing. Why are its directions random?

Some think cosmic rays originate in and are confined to our own galaxy. Even though these are not electromagnetic radiation, they are far more penetrating than the hardest gamma rays. So why should not the tired background radiation be similarly confined? **AJASC** notes that this is indeed puzzling because uniformity implies heat conduction between all parts. Yet the distances between the various parts of the Universe are greater than even light could cover in the entire age of our Universe.

I.   Fate of stellar radiation

Consider another question. What happens to all the energy radiated by all the stars in the Universe? Photons from our Sun, for example, strike the Earth, transferring thermal energy. Some of it is stored in vegetation. If the rest were not radiated away to the night sky, the Earth would heat up to the Sun's surface temperature. However, it is radiated, but at a lower temperature. Let's compare two methods for estimating the number of photons in our Universe.

Solar energy release

From Ch. 13, the Solar emission is now 3.852E33 ergs/s. In 5 billion years our Sun has emitted:

$$E_S = (3.852E33)(3.156E07 \text{ s/yr})(5E09 \text{ yr}), \text{ or } 6.078E50 \text{ ergs.}$$

Assume this rate is constant.

Universe energy release

To estimate the total Universe energy production thus far we must make several assumptions. One is that the life cycle of all the smaller stars is on the average like that of our Sun. To the 10-billion year life of our Sun, we arbitrarily add one billion years for its forming stage and one billion as a red giant at the end of its main productive life.

What happened in the 8 billion years before our Sun started forming? To simplify estimation, let us group all the Solar size stars into generations, each beginning at successive 1 billion year marks. The first will start forming at Universe age 0 and radiating at age 1E09 years. Our Sun will start forming as a 9th generation star at Universe age 8E09, start radiating at 9E09, and continue on to the present age of 14E09. There will be four generations that have radiated for 10E09 years each and are now dead. There will be nine more that will have radiated for 9E09 to 1E09 years. Their total radiation is equivalent to one generation radiating for 85E09 years. Our Universe is said to hold 100 billion galaxies each with 100 billion stars about like our Sun in size and age on the average. The total is 10E21 stars. Dividing this by 9 current generations gives 1.11E21 stars per generation.

The total population plus our Sun have produced:

$$E = E_S + (1.216E41 \text{ ergs/yr})(1.11E21)(85E09 \text{ yr}), \text{ or } 1.1472E73 \text{ ergs.}$$

Universe energy density

The volume of our Universe figured as a sphere is:

$$V = (4\,\pi/3)(14E09)^3 = 1.149E31 \text{ cubic LY,}$$

or $9.734E84$ cm^3. Assuming the energy gets uniformly distributed, the energy density is:

$$E_T/V = 1.147E73 \: / \: V = 1.179E{-}12 \text{ ergs/cm}^3.$$

Photon characteristics

Let us assume all this energy ended as microwave radiation characteristic of a source at 2.75 K. A plot of intensity I vs wavelength λ calculated by Planck's Law (Exhibit 9.1, Type 5, Section B) is a curve with a peak at the point ($6.44E{-}10$ W/cm^2 per cm, 0.1054 cm wavelength). I drops off rapidly for shorter wavelengths: at 0.0300 cm, I = $1.31E{-}12$. For longer wavelengths, the rapid drop is followed by an extended tail not reaching effectively zero until 0.7 cm wavelength.

So the effective range of wavelengths is 300 to 7000 microns, the average being 3650. By $c = \lambda\nu$, its frequency is $8.21E10$ Hz. From $E = h\nu$, its average energy per photon is $5.44E{-}16$ ergs.

Expected photon density

As a crude estimate of the number of photons per unit volume expected from observed stellar output, we have:

$$N = (1.179E{-}12 \text{ ergs/cm}^3) \: / \: (5.44E{-}16),$$

or 2166 photons/cm^3, assuming no primordial photons.

Observed photon density

The literature has various estimates of the current photon/baryon (P/B) ratio. None of the references I have consulted tells how this ratio is determined.

TFE states that the number density of photons in the microwave background radiation (MBR) is known from its temperature. (How is not obvious.) Then if the P/B ratio is known, the mass in the Universe follows. Some background is offered in terms of trial and error computer simulations of the nuclear processes that might be associated with the BB to account for the

Charles H. Peterson

relative amounts of elements found in the Universe today. Validation of the P/B ratio found seems to be the agreement of the simulations with these amounts.

> **Question:** How does one count photons? Given a huge spread in energies, is a low energy photon counted as equal to a high energy photon? What about the continuing production of photons from stars? How many assumptions are built into the simulations?

However, from it and the number of baryons we can calculate a figure for the number of photons. The range for P/B is said to be 1 to 10 billion per baryon. The baryon number density at critical density is:

5.36E79 / 9.73E84 cm^3, or 5.509 per million cm^3.

The range for photon density is 5,509 to 55,090 per cm^3, or 3 to 30 times the value expected from stellar output. Although the agreement is good at the low end of the range, the two methods may be counting different things. The expected number is for total production while the P/B method may focus on only current photons.

Analysis

Has the number been constant? It is asserted that new stars are continually being formed. Others are dying.

> **Comment: BDS** estimates that at least 20 stars in our galaxy explode every year. So we have lost 100 billion stars in the productive lifetime of our Sun. Isn't it curious that as many stars have exploded as are remaining? Does this imply regulation to a steady state population?

> **Comment:** If our galaxy is 14 billion years old, some 280 billion stars in it may have exploded over its lifetime vs the current population. This continuing fireworks of exploding stars should be of considerable interest to both Science and Religion in their efforts to explain our Universe. What purpose does it serve? It could be only incidental in both or it could be intended to generate heavier elements. **BDS** also says supernovas are observed in our galaxy about once every 300 years.

The energy output estimate above is low since it omits the output of red giants, novas, supernovas, and quasars. I do not know the size and luminosity of the average red giant, but suppose its surface temperature is half that of our Sun. Its emission would be 1/16 of the Solar value per unit area. Diameter estimates range from 20 to 200 times that of our Sun. The area could then be at least 400 times that of our Sun. The radiation would be

312

400/16, or 25 times that of our Sun for their life span, taken as 1E09 years. For 4 generations with 1.11E21 stars each:

$$E_{RG} = (4.44E21)[(1.216E41 \text{ ergs/yr})(25)](1E10 \text{ yr}) = 1.349E74 \text{ ergs}.$$

The radiation was originally as red light, but is now degraded to MBR. The energy density is $1.386E{-}11$ ergs/cm^3, and the photon contribution is 23,370/cm^3.

Novas add more. At the rate of 1/year per 5 billion stars, the total number is 2.62E22 novae over 13 billion years. The explosion results in expulsion of matter and radiation, but a dense core is left. Suppose only 1% gets converted to energy:

$$E = (0.01)(2.62E22)(1.989E33 \text{ g})c^2 = 4.65E74 \text{ ergs, adding } 80,558 \text{ photons/cm}^3.$$

The total expected is then 105,700 photons/cm^3. Of course, we don't know how long it takes to degrade a high energy photon to those in MBR.

On the other hand, other estimates of the number of baryons exceed that used here by a factor of as much as 27,000. This would put the number of photons by the P/B method as high as 149 million/cm^3.

The picture is muddied up by the claim that massless neutrinos from stars can (somehow) carry off a great deal of energy, but it is never explained what becomes of them.

Interpretation

What does this mean? If we agree there is a reasonable agreement between the predicted and the observed photon densities, we can suggest another interpretation of the observed isotropic character of the 2.75 K radiation. It is due to the photons that have been emitted by all the stars and that are bouncing around in our Universe. Bouncing because, since radiation travels in straight lines, the only way they could be streaming in all directions is if they were bouncing off the "walls" of our Universe, curved or not.

So the isotropic radiation is, not proof, but support for the hypothesis that not only is our Universe finite but it also has boundaries which are most readily thought of as closed curved rather than rectilinear. Would it stimulate your imagination to know that Lemaître thought of it as a Cosmic Egg back in 1927?

Unless there is some absorption or conversion of photons, the intensity of the background 2.75 K radiation should be increasing, and there should be increases in radiation at other "temperatures" from photons that have not yet degraded.

Overall, the story of the 2.75 K background radiation appears incomplete.

Summary

1. Science proposes that our Universe began as a singularity - a point of zero dimensions - in which was contained all matter, energy <u>and</u> <u>space</u>.

   • Science does not explain where this singularity came from or why it exploded in a Big Bang. It asserts that time began with the BB. This is at least an illogical effort to put closure to Science's efforts to explain our Universe. Something had to exist before it could explode. The BB involves many other assumptions.

   • This concept rests on interpretation of one observation: the redshift of stellar spectra, which is attributed to an expansion of the Universe itself.

   • The numbers (sizes, velocities, and time intervals) are inconsistent with a point source.

2. Science agrees with Religion that our Universe had a beginning. Step by step, Science is accepting much of what the Bible tells us.

3. The expansion after the BB is full of assertions as to what happened at so many seconds after the BB. No basis is given for the assertions.

   • Science violates its own assumptions. The velocity of light c is constant, yet for the Big Bang expansion rates greater than c are allowed.

   • The uniformity in the Universe implies heat conduction between its parts at a rate greater than c.

   • Science talks of "cooling" but not about where the heat went. The red shift means the photons have lost energy. Interstellar space contains huge clouds of gas that red shift light by two mechanisms.

4. Recombination of protons and electrons is said to have occurred at one million years when the temperature was 3000 K. This conflicts with the

ionization temperature of 105,000 K. Recombination implies the starting point was matter.

- The radius of the Universe is said to have reached a million LY at this time. Proton spacing would be about 2 billion times the range of the strong force. How did they meet?

- The repulsive force between protons that are just touching is far greater than their gravitational attraction. The strong force is supposed to be 100 times greater than the electrostatic repulsion. What is it? How is this established?

5. Science proposes that the Universe, at an early stage after the BB, existed as an enormous cloud of gas, most likely hydrogen. It broke up into many smaller clouds of varying sizes. Science cannot explain how, which is another example of faith that is needed by a devotee of the religion of Science.

- If we accept the estimates of velocities of galaxies and apply the concept of Uniformitarianism and the Conservation Laws, these velocities would have carried the galaxies back to a sphere that may be as large as a billion LY in the case of the farthest galaxies, which is not a point source.

- Calculations show that gravity should have caused everything to be receding more slowly than they are. This suggests gravity was not operating the whole time. In the beginning there were only photons.

6. Photons don't have a temperature. What is calculated is their source temperature. When photons interact with matter, they leave at a lower energy level that reflects the temperature of that matter.

7. The estimates of temperatures required for stellar nuclear reactions cover a wide range.

8. Science speaks of a degenerate state of matter, which may be Gamow's crushed state of matter. It focusses on the electrons, saying they exert a pressure to oppose the gravitational contraction. It does not talk of what is happening with the equal number of much more massive protons.

- The pressure exerted by the strong nuclear force between protons (69 trillion quadrillion psi) is inconsistent with the 150 million psi said to be needed to crush atoms.

315

- Science asserts the pressure of a degenerate gas does not depend on temperature, i.e., its kinetic energy, but does not state what it does depend on.

9. Science does not explain how, in forming deuterium from two protons, a negatively charged electron is created that then forms a neutron.

10. The Laws of Conservation of Energy and of Momentum appear to be applied inconsistently in discussing particle collisions.

11. Positive and negative particles can interact in several ways. Science offers no explanation of what determines these differences.

12. Radiation from the Big Bang was surely polychromatic. Why should it all "cool down" to only single wavelength? The background microwave radiation is hailed as the survivor of the BB radiation on the basis of calculations aimed at explaining the abundance of light elements. The assumptions need to be reviewed.

- Science cannot explain why this radiation is isotropic, nor why it is omnidirectional.

- A possible alternative explanation for this radiation is that it is the residual radiation from millions of years of interactions between stellar photons and matter during which its energy, and hence its frequency, is steadily degraded.

13. Descriptions of black holes are confusing. They are small regions of enormously great gravity, so great that even light cannot escape. Some are also extremely tenuous but massive entities.

Exhibit 14.1
Expanding Universe

Model

Let us assume that all the matter now in the Universe was once gathered into a single sphere of radius R_0 centimeters and mass M grams. Let us further assume that all the matter in the cluster of galaxies we see in the direction of the constellation Hydra was on the surface of this sphere. While this may be large, it is but a tiny fraction of the total.

Next, when the sphere exploded let us assume that this portion of matter somehow retained its integrity and left the sphere at an initial velocity of v_0 cm/s. We want to know how far it travelled in time t seconds following the explosion.

Derivation of Relationship

1. The force of gravity on the surface of the sphere was $g_o = GM/R_0^2$.

2. When the cluster reached an average distance R_1, g was GM/R_1^2.

3. Over the distance R_0 to R_1, the average g is:

$$g_{av} = GM/R_0R_1.$$

4. The distance travelled is given by:

$$R_1 - R_0 = s = v_0t - (g_{av}/2)t^2$$

5. We can proceed algebraically, but it is more direct to use calculus to find s_{max} by finding the time when the rate of change of s is zero.

$$V = ds/dt = v_0 - g_{av}\, t = 0$$

6. $\quad t = v_0/g_{av}$

7. Substituting this back in 4),

$$s_{max} = v_0^2/2g_{av}$$

$$= A(R_0)(R_0 + s_{max}), \text{ where}$$

8. $\quad A = v_0^2/2GM$

9. $\quad s_{max} = \dfrac{AR_0^2}{1 - AR_0}$

The maximum value of R_0 is $1/A$, for then the denominator is zero and s_{max} becomes infinite. This simply means that as the assumed mass M is spread out more to approach $(R_0)_{max}$, its effective g value is weaker, and a galaxy such as Hydra can travel farther.

10. The change in velocity is by 5) and 6):

$$v_1 - v_0 = - g_{av}\, t = - v_0$$

and is reached in t seconds after travelling s_{max} cm.

Sphere of Packed Protons

Suppose the mass M is made up of spherical protons, each of mass density ρ_p that are contained in a densely packed sphere of radius R_0.

$$M = [(4\pi/3)R_0^3\, k_s]\,(\rho_p)$$

Spheres can be packed in more than one way, each characterized by a packing factor k_s, which is the ratio of the volume occupied by spheres to the total volume of the arrangement. For the densest normal packing for spheres of equal diameter, it is 0.74. Substituting 11) in the expression for A in 8):

$$12.\quad A = \left(\frac{v_0^2}{2G}\right)\left(\frac{3}{4\pi k_s \rho_p}\right)\left(\frac{1}{R_0^3}\right) = \frac{v_0^2}{R_0^3}B$$

$$B = 3/[(8\pi G)(k_s)(\rho_p)]$$

$$= 3/[(8\pi)(6.672E{-}08)(0.74)(1.45514)]$$

$$= 1.661E{-}08 \text{ seconds}^2.$$

substituting in 9):

$$13.\quad s_{max} = \frac{\dfrac{Bv_0^2}{R_0}}{1 - \dfrac{Bv_0^2}{R_0^2}}$$

From 13), we find there is a minimum value of R_0^2 for the case of a portion of the material expelled from the surface of the sphere at the velocity of light:

14. $(R_0^2)_{min} = Bc^2 = 1.493E13$

15. $(R_0)_{min} = 38.64$ km,

very much smaller than that for dense packing. Is this possible? As R_0 decreases toward $(R_0)_{min}$, the numerator increases, but the denominator decreases at an ever faster rate, so s approaches infinity even faster. If R_0

could go below this minimum value, s would go negative. In the real world, no one knows what happens if one could get to infinity. Is there a negative world beyond it?

The force of gravity at the surface of the minimum sphere of gas by 1) is:

16. g $= (6.672E{-}08)(8.96E55)/(3.864E06 \text{ cm})^2 = 400.4E33 \text{ cm/s}^2$,

or 410 quadrillion quadrillion times that on Earth. A 1-second application of g would induce a velocity toward the sphere of 420 quadrillion LY/s. Since light travels at 1 LY per <u>year</u>, we must be dealing with a black hole.

However, since both the gravitational force and the Coulomb force are inverse square laws, as R decreases the Coulomb force will always be greater than the gravity force. In Ch. 11, we noted that the strong force decreases as the seventh power of distance. So the protons can't escape. What happens to them if the sphere of densely packed protons can be compressed to the limit of Equation 16?

Chapter 15. Fossils and Evolution

<u>Review</u>

We have been on a long journey to the outer limits of our Universe. We have peered into the atom. We have been seeking God. What have we found? My perception is that Science has not permitted us to advance one inch toward answers to the questions I was able to ask in my youth. Science has not <u>explained</u> anything truly fundamental, like what is:

Space	Time		
Charge	Electricity	Magnetism	
Matter	Gravity	Radiation	Energy

We have also found that:

• The theories rest on many assumptions, including those of point masses and point charges that inherently prevent exploration and understanding of the ultimate structure and nature of these entities;

• The experimental supports for some of the important hypotheses of Science are microscopic or entirely lacking;

• Science selectively ignores its own principles and proofs: Ockham, Gödel, Conservation of Energy.

• It applies the hypotheses inconsistently.

• Its investigations merely uncover greater mysteries than they sought to explain.

• It proposes increasingly bizarre conceptions that we are to accept, not on proof, but on faith.

Science has been fabulously successful in <u>describing</u> many of the ways in which our Universe <u>behaves</u>. But it can do no more than this. It and its gadgets give us the false hope that we can control our Universe. It describes the What, the When, and often the How. It has been <u>totally</u> unable to explain the Why or the Who. Since it can't, Science tries to persuade us there is no Because and asks us believe its creeds in which there is no Who.

And it is like a little boy finding firecrackers, discovering they explode, not knowing how or why they came to be, but choosing to believe they were always there or created themselves. We still don't know a) if our Universe is closed or open, or b) is there a God?

Science thus offers us "explanations" for the things and events in our Universe that amount to a secular religion for us to live by. Its strategy is to direct our attention away from the things it can't explain by rarely talking or writing about them.

> **Comment:** Religion can therefore apply the same logic when it finds resistance to the idea of God: we do not need to understand it any more completely before we agree to act on it.

This religion of Science appears to rest on at least the following dogmas, creeds and beliefs:

- The Dogma of Uniformitarianism (James Hutton, 18th century) preached by Charles Lyell (19th century). It claims natural laws and processes aren't changed by time or distance and so can be used to explain the origins and the development of our Universe and all the life forms in it (**TCOE**).

- The Dogma of Evolution that is completely rejected by many prominent scientists and even former evolutionists, yet our laggard authorities insist on forcing it on our children in our schools.

- The Doctrine of the Lorentz Contraction, which underlies the Dogma of Relativity.

- The Doctrine of a constant speed of light, a shallow analogy to the constancy of the God of Light.

- The existence of many "Natural Laws", whose origin Science can't explain.

- The Book of Thermodynamics.

Science has thus closed its mind to the reality that what it offers is just another belief system. Its several articles include the faith that its basic tenets can be relied on. Until they are replaced. . . .

> **Comment:** Further, high energy physics has become a black art: its methods and the pronouncements of its high priests are incomprehensible to most.

And Science refuses to address our <u>universal</u> desire, urge, craving for something to believe in, some statement of meaning and purpose, some explanation of WHY.

Question: WHY do we ask why? Why are we not satisfied with Science's answer that there is no Because?

Our craving for answers extends to devising cults complete with rituals and incantations to invoke the Unknown. I have begun to suspect that what really drives humans is fear, a deep-seated consciousness of a fear of a Power that somehow brought into existence all the things we see, and probe, and measure. That fear is so intolerable that we do everything possible to minimize its impact on our psyches.

- We fill our hours with busy-ness to keep it out our conscious minds.

- We rationalize and redefine our terms and concepts to construct a mental world that does not permit a God to enter.

- We ridicule the idea of God to bring it down to our puny level of power. We seek to persuade or coerce others to join us in this ridicule.

- We hide from Nature in our buildings and cities and rarely see a sunset or a rosy-fingered dawn spread its matchless splendor across miles of sky.

- We light up the night, blotting out the light of the stars so that we can pretend their wondrous panorama doesn't exist.

Yet it is a Power we infer because the basic clue is all around us: <u>everything</u> has been caused by something. Billions of everythings. From cosmic to infinitesimal.

- The sand on the beach came from rocks due to temperature changes and the action of water.

- The temperature changes because of the change in the seasons that are due to the tilt of the Earth's axis and the orbiting of the Earth.

- The water came from hydrogen reacting with oxygen which came from hydrogen via nuclear reactions in stars like our Sun.

- The stars came from hydrogen.

- The hydrogen came from . . .?

Permit me a couple of assertions:

- There is not one thing that does not have an ultimate cause.

- In every case, that cause is outside of the thing that is caused.

Science vigorously pursues its experiments, thus demonstrating its unshakable faith in its concept of cause and effect relationships. Yet its tunnel vision is shown by its refusal to accept the fact that <u>none</u> of these relationships disprove the existence of a Creator, any more than fully understanding the mechanisms of a motor car refutes the existence of a manufacturer. Read **Job 38**. Here are three samples:

38: 4 "Where wast thou when I (God) laid the foundations of the Earth? Declare if thou hast understanding."

38:32 "Canst thou bring forth Mazzaroth (the zodiac) in his season? or canst thou guide (the star) Arcturus with his sons?"

Comment: This reminds us our powers are puny relative to the enormous forces in our Universe.

38:36 "Who hath put wisdom in the inward parts?"

Comment: The ancients already had the understanding that the workings of our bodies were indeed complex. <u>We</u> take them for granted, which means we don't want to bother our heads or disturb our psyches by contemplating how they came about.

Science seems to dissent from Religion on but one basic point: rejection of the concept of an Intelligent Creator, a personal God. But in its unceasing, unheeding, unbridled quest for material knowledge it callously offers us only a heartless, helpless, hopeless nothingness. For such is the nature of its gods.

Psalms 115:4-8 "Their idols are. . .the work of their own hands. They have mouths, but they speak not; eyes. . .but they see not; . . .ears but they hear not; . . .They that make them are like unto them; so is every one that trusteth in them."

It is true that both Science and Religion can take the same stand on the ultimate question: where did our Universe come from? The answer: it or God always existed. But across the centuries comes the same question that has been in billions of tormented souls and anxious minds:

Job 11: 7 "Canst thou by searching find out God?"

Question: From whence cometh the thought, the desire, the drive, to search for God? If we are only a collection of atoms, why do we care?

Science is succeeding in showing there is a decreasing probability that it will ever provide a definitive no to our primary question.

- Its theories of the structure of matter deal with dimensions so infinitesimally small that Science itself despairs of assigning any meaning to the possibility of even smaller sizes. Further, the apparatus needed to explore the frontiers is prohibitively expensive.

- The investigations into the remote reaches of our Universe have run up against our "horizon": light from greater distances, if it exists, just has not had time to reach us.

We are still left with the choice between the approach of Science and that of Religion. We might have hoped that we would have a firmer basis on which to make this choice. It might be argued that this review has only been able to show that Science can no more prove its position than Religion can. But perhaps you, too, will reject the proposition that we are food processing systems whose only role is to indulge our senses and to convert everything we ingest to sewage.

What can we say to tilt the balance? Recall that the purpose of this book is to answer only one question: Is there a God? It is not concerned with what we believe about that God or how we behave toward Him.

Perhaps we should look at another pair of mysteries: the origin of life on Earth and its diversification. Let's review some of what is known of geological strata and geological time. We will skim through systems for classifying life forms as a preparation for considering the nature of fossils in the oldest strata. Finally we will examine some of the ideas of Evolution. Our review is not intended to make us experts in these areas. We merely wish to understand what the experts are saying and why. Let us continue to be alert to the number of unanswered questions.

Paleontology

A. Hydrological Cycle

We have talked a little about the early Earth when it was a cooling mass of molten stone and iron. Eventually the Earth would be cool enough to permit water to condense as rain. This began the amazing hydrological cycle that also started soil formation by disintegrating the rocks through freeze-thaw cycles to form countless billions of small solid particles. It would also leach alkali elements from rocks, as soluble hydroxides, chlorides, sulfates, and phosphates.

The residue of insoluble solid particles of oxides of silicon, aluminum, magnesium, iron, etc. would be carried by wind and water to collect in low-lying areas as sediments and eventually get compacted into sedimentary rocks like sandstones and shales. Later, chemical reactions occurred to form metamorphic rocks.

> **Comment:** This is more elegant and fitting than if God personally pounded the rocks with a sledgehammer for a few billion of our years. Yet, if God can by speaking bring the Universe into existence, why didn't He do it in one step?

B. Strata

   1. Description

Geologists and paleontologists tell us of three observations about the Earth's crust, which is about 35 km thick. This is 1/200th of the Earth's radius. It is like the skin of an apple. Or a scum 1 mm thick floating on 20 cm of water.

In various places, we can see the crust consists of strata (layers) of different thicknesses, materials, and particle size distributions. The strata are made of sedimentary rocks and/or metamorphic rocks. They rest on igneous rocks. In these strata are found evidences of long dead life forms, called fossils.

The thickness of the layers is from less than a centimeter to more than 30 meters. Each layer may be made up of many thinner layers. It is not clear how these are declared as parts of a single layer. Most must have accumulated under water: sea beds, lake beds, river plains. **TOG** doubts if sedimentary strata were ever more than 20,000 feet thick. To explain such beds, the sea bottom must have subsided. Yet **TOG** accepts a thickness of over 94,000 feet (!) for the Archeozoic calcareous Grenville strata in Ontario.

   2. Interpretation

The layers are said to represent different periods in Earth's geologic history. They are due to recurrent flooding by changing ocean levels from changes in the elevation of the sea bottom or thawing and refreezing of the polar ice caps. The totality of layers is the geologic column, although the entire column is not found anywhere on Earth.

Often there are discontinuities between strata. Many of these are interpreted as lost erosion intervals in the geological record. **TOG** says major breaks occurred in the middle of the Proterozoic, and between every one of the eras. **DE** notes that at various places in the world, rocks of every geologic period, not just the Cambrian, lie directly on the basement rocks. This is as difficult to explain as the Cambrian explosion.

> **Comment:** Perhaps it is not strange. First, there must be a source of sediment. Land-based sediments accumulated from particles of silica (sand) and silicates (clays) from weathering of rocks. Ocean-based sediments accumulated from the billions of skeletons of tiny plant and animal life forms over thousands of years, as well as land sediments washed down to the oceans. When snow and rain were heavy, the rate of transfer would be high. When light, thinner strata would be formed. If the rate at which life forms died was constant, thin strata would show a higher concentration of fossils. But maybe there never were any strata deposited in certain periods of time, rather than that they eroded away. This would also explain why strata of different periods can lie on a given older stratum.

3.   Formation

How did each of several layers get deposited without disturbing the previous one? How did they get compacted? Very fine-grained clays do cohere, but not to the extent of forming stone. Water pressure is a possibility. On the Continental Shelf, the water depth may be 600 feet. This would exert a pressure of 260 psi.

Consolidation may have involved chemical cementing via carbonates and silicates because water dissolves carbon dioxide from the air making carbonic acid and attacks the silicate structures in rocks directly. But where did the carbon dioxide come from? Was there an initial inventory in the atmosphere?

C.   Biblical flood

1.   Creationist argument

Creationists argue that the geologic column is a fiction since it does not exist anywhere in its entirety and that the layers were all due to a single worldwide Flood less than 10,000 years ago.

> **Question:** Why would there be layers? The Flood was over in a year. Bones from humans and land animals would have sunk to the bottom of the Flood in a tangle. They would not have been distributed across many

vertical feet of sediment. It is not credible that the many layers came from a <u>single</u> flood.

2.   Mathematics of the Flood

The notion of a worldwide Biblical Flood is extremely puzzling. So let's pause to examine it from the view of the laws operating today. How great was the Flood?

Genesis 7:20 "Fifteen cubits (22.5 feet) upward did the waters prevail; and the mountains were covered."

Mt. Ararat is 16,945 feet high. The elevation of the surface of the Flood was then 16,967 feet.

Calculation: The volume of water added above sea level is the difference between the volumes of two spheres, one of radius 3963 miles, or 20.924640 million feet, and the other of radius 20.941607 million feet. This works out to 93.4E18 cubic feet. At 62.4 pounds/cubic foot, we have 2.64E24 grams.

- Did the water come from the atmosphere?

Handbooks show about 5E21 grams of air in the atmosphere. So the Flood water amounted to 2.64E24/5E21, or 529 grams per gram of air. However, the maximum is at 100% relative humidity, and depends on the temperature. At 70°F and 100% R.H., the air can hold only 0.015 g water per gram of air. Adding 529 g of water to every g of air would greatly increase the atmospheric pressure. The air column pressing on each square inch of the Earth's surface weighs 14.7 pounds. Adding the water increases the pressure by 529 x 14.7/2000, or almost 4 tons/in^2. Noah and his family would be rather flat.

The water vapor would also create a large greenhouse effect: it emits less energy as infrared than it absorbs. Noah et al. would have been cooked.

- Perhaps it came from the oceans.

The Flood may have been associated with the breakup of Pangaea. Perhaps an enormous meteorite struck the Earth, pushing the continental fragments down into the underlying magma as well as disrupting the weather patterns. The breakup could also expose large amounts of magma to sea water. Enormous quantities of water would vaporize, the clouds would drift to cooler areas to fall as rain.

- Next, it took 10 months for the waters to recede so the tops of the mountains could be seen.

   Question: Where did the water go? To evaporate it would require an enormous amount of heat. And it could not run off, because there was no lower place for it to flow to. It is simpler to assume the continental fragments rose finally to about the original hydrostatic equilibrium levels.

Many discount the Great Flood as a myth. Apparently some humans from all over the world did survive this catastrophe (and all other floods) and preserved the memory of it in their folklore. This suggests it was more than an ordinary local flood.

North America seems to have been flooded at least sixteen times since the Proterozoic Era, but hardly at all in the last 60 million years. Paleontologists suggest that humans were around for at least 60 million years.

Some Creationists hold that dinosaurs were contemporary with man. It is not impossible that there was some overlap between early man and late dinosaurs at about 70 million years ago. But how certain is the date?

D. Nature of Fossils

Exactly what do we see when we look at these strata? Textbooks usually omit such details. A fossil is any trace of an animal or plant of a geological period, including skeletons, shells, organic material, degradation products, footprints, imprints and even deposits bearing limestone or carbon. These materials provide clues as to conditions on Earth in ages past.

It is crucial to understand how a life form becomes a fossil. A dead animal must be buried quickly and deeply in some sort of solid material. This excludes oxygen, which thereby inhibits bacterial/mold action and chemical action from substances dissolved in the water. The sediment that buried it must then get compacted to form the strata we find. Most fossils are those of marine creatures. Most preserve the original material of the life form, but silicates or carbonates may either replace the calcium phosphates in bones or deposit on them to preserve shapes.

   Question: Is there any evidence of long term leaching of the bones or shells by sea water?

Intact specimens are rare. Notable exceptions are the mammoths frozen in Arctic tundra or the remains found in the La Brea asphalt pits.

Question: How long does it take to bury a large specimen? Deposition of dust or sediment should take many decades. If this is the mechanism for burial, why did the dead specimen remain intact for so long? Fossil trees are found spanning several strata, and hence millions of years (**TCOE**). Why didn't the tops rot away?

E. Age of Strata

1. Geological indications

The next problem is dating the strata. Earlier it was assumed that if Layer B rests on Layer A, it is younger than Layer A. If A is found resting on B, sometimes this can be explained by the nature of the strata and by earth movements. Actual ages were estimated by studies of erosion, sedimentation, and rock crystallization rates. These surely had to be very rough, and very uncertain.

Then it was observed that some of the fossils found were very simple animals while others were quite complex. The "simpler" forms often were found in the lower, hence older, strata. This suggested that there had been a continuous progression of life forms from the simple to the complex.

Comment: Somewhere, the logical error was made of assuming that, because fossil B came after fossil A, B descended from A. Thus:

- All children are descendants of their parents.
- Therefore a particular child comes into existence after its parents.
- Fossil B comes into existence after Fossil A.
- Therefore, Fossil B is a descendant of Fossil A.

Accept this logic and you accept the Dogma of Evolution. Today, a child always develops into an adult like its parents, but Science says Uniformitarianism does not apply to fossils.

Question: How was it decided that millions of years instead of thousands were involved? The popular literature also does not describe exactly how the fossils were spaced in the strata and hence in time.

2. Radionuclide dating

Radionuclide dating is the only method that might give quantitative estimates of dates. Besides the question of whether the decay rate was constant discussed in Exhibit 12.1, there is another possible error. Any exchange of lead or uranium with the surroundings would make the final Pb/U ratio either too high or too low, and the calculated age would be too old or too young.

However, dating is not based on just one sample or one locality. While some samples may be suspect, it is unreasonable to say they are all suspect. The results have to be consistent with other observations such as those of geology. The process of interpretation is like getting a majority opinion. The alternative is no opinion, an abdication of our power to reason, which is contrary to our human nature.

But then we read in **EST**, Stratigraphic nomenclature, a most startling confession: most sedimentary rocks do not contain measurable radioactive materials, so this method is impractical! Dating of such strata thus depends on the fossils found in them and the assumption that there has been a succession of ever more complex life forms. However, radiometric dating is possible for igneous rocks and some metamorphic rocks.

The oldest ages found are about 3800 million years. So no life can be dated older than this. It happens to be about the beginning of the Archeozoic (Table 15.1). There may be older rocks not yet found.

> **Question:** But how <u>were</u> the fossils dated? Clearly, we need to look at the original data.

F.   Geological eras

Earth's geological history is now divided into six eras. Each is represented in nature by many strata, which indicate many periods demarcated by some large scale change, as flooding, warming, cooling, etc. The figures in Table 15.1 are the estimated times when these eras began millions of years ago and are mainly from **RHCD**, Era. I have not examined the original data so I do not know what the uncertainty in the dates is.

Each era is divided into periods, only some of which are listed. Of particular interest is the Cambrian Period, beginning about 600 million years ago. While your attention is probably attracted to times when particular life forms appeared, the early times when nothing seems to be happening are really more significant for our review. **TOG, DE,** and **EST** give different Periods for the appearance of some of the major life forms. **DE** and **EST** put the early mammals in the Jurassic Period; **TOG** says Triassic; **RHCD** says Paleogene.

Table 15.1 Geologic Eras

Era	Period	Began[a]	Life Forms
Cenozoic			
Late	Quaternary	1	Widespread glaciation; man
Early	Paleogene	60	Modern mammals
Mesozoic			
Late	Cretaceous	135	Giant reptiles extinct; flowering plants
Middle	Jurassic	180	Flying reptiles; birds
Early	Triassic	220	Marine reptiles; dinosaurs
Paleozoic			
Late	Carboniferous	350	Winged insects; conifers: bare seeds
Middle	Devonian	400	Amphibians, lungfish/
	Silurian	440	Air-breathing animals
Early	Ordovician	500	Armored fishes
	Cambrian	600	Marine invertebrates; green algae
Proterozoic		1500	Bacteria; blue-green algae
Archeozoic		3000	Fossils unknown; unicellular life?
Azoic		5000	No life possible

[a]Millions of years ago

G. Classification systems for organisms

1. General

Way back in grade school we were taught that there are three great Kingdoms: Animal, Vegetable, and Mineral. Animals can move around on their own, plants cannot, and minerals are not alive. Also, plants are green and have roots. These distinctions were made on the basis of the large life forms we saw around us.

2. Current systems of classification

The microscope revealed the existence of microbes. They did not fit the purely Animal or Plant Kingdoms. Three new Kingdoms were defined to deal with bacteria, algae, protozoa, and fungi; 13 more were proposed.

EST, Taxonomic categories, describes four systems of classification. One, the evolutionary, attempts to show ancestor-descendant relationships. However, it was found that life forms could not be placed into categories in which the members had characteristics assigned only to that category. Even Lamarck said no sharp distinctions existed for classifications. The cladistic assumes ancestors cannot be identified, and focusses on branching relationships.

A biological category is thus merely a human effort to group life forms according to similarities and differences. These groupings are then ranked from simple to complex by the researcher's judgment. One category is based on hard fact. If two different life forms can mate and produce fertile offspring, they are of the same Species. If not, they are of two different Species. Two or more Species make a Genus, etc.

> **Comment:** Recognizing this reveals a crucial point in the Evolution/Creation controversy. Something had to intervene repeatedly in the endless succession of generations to produce new Species that could not mate with their contemporaries.

Important classification criteria now include:

- Does it have cells with nuclei?
- Is it single-celled or multi-celled?
- Does it make its own food?
- Does it reproduce by an embryo?

PCAI focusses on invertebrate animals using a classification system emphasizing body structure rather than specific organs. Vertebrates have internal backbones while invertebrates have none at all, but may have shells. **EST** recognizes 28 phyla in Animalia, whereas **PCAI** lists only 16. They agree on nine. Four more are treated by **PCAI** as Subphyla, ten others as Classes, and does not even mention the other five.

> **Comment:** If the experts can't even agree on what is a Phylum, a Subphylum or a Class, how can any pronouncements be made as to which is related to what?

In the Division Deuterostomia in **PCAI** is the Phylum Chordata whose members have at least a column of cells called a notochord made of a gelatinous unsegmented material that suggests a primitive backbone. Tucked away in Chordata is a Subphylum Vertebrata, in which all the members have true backbones. It includes fish, amphibians, reptiles, birds, and mammals. That's us.

Comment: Our perspective on life is warped by the attention we give to the large life forms, ignoring the tens of thousands of other Species. Since each Phylum has at least one Subphylum, Vertebrata by **EST** is about 3% of the total. The Class Mammalia is only one of five in Vertebrata.

3. Composite classification

Table 15.2 illustrates how the two systems might be meshed. It preserves the eight classical levels. But notice how many insertions are needed in the form of sub-, super-, and infra- levels. Consider what this means as to the number of "random" events needed to produce so many differences.

The Primates include man; the apes; monkeys; and others. Hominoidea include the great apes (chimpanzees, gorillas, and orangutans) and man. Some place the apes in Pongidae and only humans in Hominidae. The Table compares a human and a chimpanzee.

Question: Does this necessarily mean these are descended from a common ancestor? Or does it mean only that there are observable similarities (facts) and the rest is speculation?

We can go through the same exercise with motor cars. In America, for the Superphylum Wheeled Land Vehicles there are three surviving Phyla: General Motors, Ford, and Chrysler. All are built on animal prototypes with four wheels (legs), headlights (eyes), a horn (mouth), an engine (heart), an air intake (lungs), cooling systems (blood, perspiration), and an exhaust system. They do not have reproductive systems.

Any car buff could identify the Subphyla easily. In General Motors, there are Chevrolet/Buick/Pontiac/Oldsmobile/Cadillac. Then there are Classes: passenger cars, trucks, vans. And Subclasses: 2-door/4-door/hatchback. And we have done God one better by labelling the model years. The classifications would have the following meanings:

Classification	Having
Metazoa	Many components
Enterozoa	A fuel processing system
Bilateria	Bilateral symmetry
Deuterostomia	Separate fuel intake and exhaust

Question: If you were an alien visitor to Earth, could you avoid noticing the similarities in characteristics (designs) of wheel sizes, the number of seats, trunk designs, windows, roof designs, etc. Surely they have evolved one from the other.

Does anyone really believe there was no designer who made conscious decisions on these designs?

H. Kinds of Fossils

We'd like to know the probable sequence of life forms Evolution postulates. An increasing degree of complexity would be something like the following:

- Single-celled with no nucleus
- Single-celled with a nucleus
- Multicelled with nuclei

What is actually found in the geologic record?

1. Archeozoic fossils

The bedrock is igneous and has no fossils. Not surprising. No life forms can survive at 6000 K. Nor are fossils found in the overlying strata. But it is claimed that calcium carbonate deposits mean life was present 3500 million years ago. Dating methods are not given. The rationale is Uniformitarian: algae today lay down such deposits. It is also claimed that graphite deposits are due to bacterial metabolism as the only known natural mechanism for reducing carbon dioxide to carbon.

Question: Some bacteria convert CO_2 to <u>compounds</u> of carbon. Others extract <u>oxygen</u> atoms from such compounds for their life processes, thus forming hydrocarbons (petroleum). What basis is there for postulating life forms that extract <u>hydrogen</u> atoms to leave graphite? There is no reason to believe that <u>all</u> carbon originally was oxidized. How old are diamonds? Graphite deposits could be primordial carbon from nuclear processes in stars.

From **EST** (Algae, Cyanobacteria, Bacterial metabolism), we would conclude that the earliest life forms were blue-green bacteria, i.e., single-celled, mobile organisms without nuclei, and capable of making their own food from inorganics like water and carbon dioxide by photosynthesis. They appeared between 2000 and 3500 million years ago. But if an alga (a plant) is a lower form of life than an animal (a bacterium), Evolution is going in the wrong direction.

Table 15.2
Composite Classification System

		Man	Chimpanzee
Kingdom	Animalia		
Subkingdom	Metazoa		
Superdivision	Enterozoa		
Division	Bilateria		
Subdivision	Deuterostomia		
Superphylum	-		
Phylum	Chordata		
Subphylum	Vertebrata		
Superclass	-		
Class	Mammalia		
Subclass	Theria[1]		
Infraclass	Eutheria[2]		
Legion	-		
Sublegion	-		
Superorder	-		
Order	Primates		
Hyporder	Anthropoidea		
Suborder	-		
Infraorder	Catarrhini		
Superfamily	Hominoidea[3]		
Family		Hominidae[4]	Pongidae
Subfamily		Homininae	Ponginae
Tribe		-	-
Genus		Homo	Pan
Subgenus		-	-
Species		sapiens	troglodytes
Subspecies		-	-

[1]Means beasts vs earliest beasts and other beasts
[2]Means placental mammals
[3]Tailless; have appendices
[4]Only human forms; some workers include Pongidae.

Algae absorb certain B-vitamins from the water they grow in, their source being the metabolic processes of bacteria! So they could not have been first unless the early forms did not need vitamins. We also learn that all living things require not only enzymes but also coenzymes. Some bacteria cannot synthesize them because they have lost the ability to make the necessary B-vitamins. So the earliest bacteria had abilities needed for survival that were lost in later generations. Evolution is headed wrong again.

The cyanobacteria changed the Earth's atmosphere from reducing to oxidizing.

> **Question:** Is it claimed they could survive in both atmospheres? The Earth's atmosphere must have had a significant amount of carbon dioxide and water vapor to serve as food for these bacteria. This is a highly oxidized atmosphere. It fits with Venus and Mars having carbon dioxide atmospheres.

2. Proterozoic fossils

A few fossils of primitive marine invertebrates are dated to the Proterozoic Period. An example is Radiolaria (a species of Protozoa with siliceous shells).

EST, Plant evolution, says autotrophic eukaryotes (plant cells with nuclei) show up at the start of the Proterozoic. This enabled development of sexual reproduction in plants.

The late Proterozoic had only simple life forms. However, there are the Australian Ediacaran deposits of soft-bodied metazoans (multicellular animals). They are not considered ancestors of hard-bodied Cambrian fossils. They are widely distributed geographically, but are not found in Cambrian strata so they must have ended in a mass extinction. Some deposits are only 30 vertical feet below the oldest Cambrian fossils, while others are thousands of feet lower.

> **Question:** What were the single-celled precursors of these metazoans? This also assumes that the strata separating the fossil locations is Proterozoic.

3. Cambrian fossils

Exactly what does mark the beginning of the Cambrian? A boundary is where the rock type "suddenly" changes or where different fossils appear. If the Proterozoic ended by a climate change, the time till the first Cambrian fossils might have been indefinitely long.

TOG says in most places where Cambrian rocks rest on Proterozoic or earlier rocks a marked unconformity occurs that represents a loss of geological record. However, there are places where no loss of record is evident. There is also no way of estimating how much erosion occurred, if any.

Comment: There are 900 million years in the Proterozoic and another 1500 million in the Archeozoic to speculate about vs the 600 million with fossils. We can say either there were no Pre-Cambrian fossils or agree with **TOG** on the loss of geologic record. We could also choose to believe in Creation.

An astounding phenomenon is the Cambrian Explosion. "Suddenly" the strata have many diverse invertebrates.

Comment: Science must postulate "natural" mechanisms whereby all these diverse life forms came into existence in a geologically short time. The fossils appear over a depth that corresponds to tens of millions of years (**DE**). Or 100 million years (**RHCD**). Hardly sudden.

Comment: The trilobite period from appearance to extinction covers some 400 million years. If one finds a trilobite x feet down in some sediment, how would that establish the age of the sediment?

No vertebrates are found in the Cambrian. Evolutionists and Creationists at least agree man came later. The fossils include representatives of nine of the fifteen modern <u>invertebrate</u> phyla, as shown in Table 15.3 in bold. From **EST** and **PCAI**, we can add two more (11 of 15 is 70%). This is an amazing percentage evolutionists can't explain. More importantly, the life forms are complex and should not be found so early in the geologic column. Two more Phyla show up in the next period, the Ordovician.

The shells of most Cambrian fossils are built from calcium phosphate rather than calcium carbonate, like some later life forms. This is a shift from the Radiolarian siliceous shells. Why? How?

Question: What did they eat? Primitive life forms were bottom feeders. Later the ability to filter plankton out of sea water appeared. Brachiopods ate algae, a form of phytoplankton. Animal forms are zooplankton, such as Radiolaria and Foraminifera.

Question: Corals were widespread, even in the polar regions, showing that all the oceans were warm. How was this possible?

Answer: At first, the entire surface of the Earth was hot but gradually cooling as heat from the interior was lost to space. As the crust thickened by crystallization from magma, its thermal insulating effect increased, so the surface temperature would continue to decrease.

There were about 2000 varieties of trilobites. Their lengths ranged from a half-inch to two feet. Their bodies had three lobes, and the thorax was articulated. Each had a shell, up to 29 pairs of many-jointed legs, usually two eyes and were sexed. The eyes were made of calcite, a crystalline form of calcium carbonate, and were usually compound with up to 15,000 lenses (**TOG**). Rather complex, no?

> **Comment:** Trilobites dominated the early Cambrian and were in the highest Division of invertebrates, so they could <u>not</u> have originated then (**TOG**).

> **Comment:** "Highest" is not defined; it implies "most complex". "Could not have originated" means **TOG** writes as an Evolutionist. Maybe God used some variant of evolution but made all the precursors disappear from the record. You know, like the bricklayers take down their scaffolds and we don't have any clue as to how the brick walls were put up.

> **Question:** Many life forms appeared only to become extinct. Why did God allow this? In any case, there is the impression of a powerful force operating continuously to develop more complex species.

I.   Biological development of animals

In this section, we look at some body structures and their increasing complexity. The references simply present the descriptions, but do not explain how the observed changes occurred. So the most important word in these descriptions may be the implied "somehow".

1.   Digestive

Protozoa have a vacuole (a cavity) for digestion and excretion. The specialized chemical and enzymatic processes of digestion were thus isolated so the protozoan would not digest itself. Food is engulfed and digested anywhere in the body, or taken in through the cell walls and directed to the vacuole, somehow.

In sponges, the digestive vacuoles are in mobile cells called phagocytes that are able to produce acid and alkaline conditions for digestion. In Coelenterata, the phagocytes have (somehow) formed into a single digestive organ that has a true mouth. **PCAI** identifies the coelenterates as the first predators. They usually have some form of tentacles to seize prey.

Table 15.3 Cambrian Invertebrate Phyla (**DE**)

Protozoa		
	Protozoa	Radiolaria (amoeba-like)
Metazoa		
Enantiozoa		
	Porifera	Sponges
Enterozoa		
Radialia		
	Coelenterata	Jellyfish; corals
Bilateria		
Protostomia		
	Scolecida	Flatworms; roundworms
	Mollusca	Snails; mussels; clams; squid
	Articulata	Segmented worms; trilobites
	Brachiopods	Brachiopods (may be Proterozoic)
	Crustacea	Ostracods (bivalves)
Deuterostomia		
	Pogonophora	Worm-like tube dwellers
	Echinodermata	Starfish
	Chordata	Lancelets (fish-like)

Comment: The simpler forms of life are also predators if they can't make their own food: they must ingest it from the environment. Many have cilia to move currents of water along with whatever is in that water toward their "mouths". Perhaps **PCAI** sees this as passive activity and predation as actively chasing food. An animal thus differs from a plant in not making its own food. The simplest animals live by eating plants. Complex animals kill and eat other animals as in the post-Eden world.

Comment: We make a profound observation: death was in the world long before man, and therefore should not be attributed to the sin of Adam and Eve. Death is an inherent part of animal life. Theologians can argue as to God's intent. Was this an error? An unforeseen consequence? A tolerated side effect? An intended effect?

2. Neural

Sponges have no neural system. Their cells respond individually to stimuli, but there seems to be slow communication of stimuli between cells.

All Enterozoa, beginning with the Coelenterata, have a definite neural system and muscle systems for predation. They have neurons and synapses, and produce complex chemicals like acetylcholine or noradrenalin.

> **Comment:** Somehow they acquired the blueprint for making these automatically. And how did they get the control mechanisms?

> **Comment: PCAI** makes a profound inference here: the structures arose because of the transition from passive ingestion of suspended particles of food to active predation. But which came first: the structures or the predation?

A predator needed a swift-acting muscle system to capture a mobile prey. Thus, something in generations of medusae, jellyfish, polyps and corals gazed longingly at the food passing by and was thereby stimulated to grow the apparatus to capture it.

> **Comment:** This sounds like Lamarckism + Lysenkoism: characteristics can be acquired and then inherited. Both have been discredited. An alternative version is that a living organism has an inherent capability to produce descendants with superior sensing ability and swifter reaction times.

> But how? Science has no explanation. We can exercise our muscles to greater size and strength, but no one has ever grown an entirely new muscle. We do have a third possibility called Creation: the capability was inherent in the force called life. But suddenly I saw a scripture in a different light.

> **Matthew 6:17** "Which of you by taking thought can add one cubit to his stature?

> **Comment:** Try growing wings because you strongly want to fly like a bird.

3. Sexual

Protozoa are asexual. They reproduce by mitosis, buds or spores.

Sponges (Porifera) are either hermaphroditic or unisexual. They produce separate ova and sperm cells in any part of their bodies. In some, these are discharged into their water environment where they unite and grow to adult forms. Fish use this method today.

> **Comment:** Even the most primitive Metazoa are thus consistent with **Genesis** in that God created male and female of all species, except that mitosis in Protozoa is not recognized.

Coelenterata are somewhat more advanced genitally. Sperm and ova are separate. Sperm are expelled and then find their way into the body of the same or a different member of the species to fertilize the ova.

Worms (Scolecida) have actual genital apparatus. Most flatworms are hermaphrodites. The most primitive version is immature sperm and ova in the surface tissues. There are no ducts or copulatory organs. Mature sperm enter the gut and migrate to the mouth. The worm then transfers packets of sperm to the skin of another member of the species. They penetrate its body and fertilize mature ova. Others in this Phylum develop various types of male copulatory organs.

> **Comment:** It may be of interest to reflect that humans appear to preserve this behavior as kissing, although the function of reproduction has been transferred to other organs.

> **Conclusion:** Multiple highly specialized biological systems were already in place at least 500 million years ago. No obvious precursors.

J. Biological development of plants

1. Thallophytes

These were the earliest members of Plantae (algae, fungi and lichens). They have no true leaves, stem, or root. Algae have chlorophyll; fungi do not; and a symbiotic union of these is a lichen. Most algae are sexual; some reproduce by asexual spores.

> **Comment:** If you think having chlorophyll is the most important characteristic of a plant, then you will call a bacterium a plant. If you think mobility is most important, then it is an animal. Blue-green bacteria are thus "plantimals" or "aniplants". They move by ejection of mucilage (jet-propelled); pseudopods (like an amoeba); undulations or peristalsis; and flagella. These are in the earliest life forms.

341

Algae are the only plants found for some 3100 million years until the Devonian Period (**EST**, Plant evolution). The early plants were not woody, so nothing survived except possibly breakdown products of chlorophyll.

> **Comment:** Algae are not just a green slime: they generate a large fraction of our oxygen. Also, if Evolution is correct, they are our remote ancestors. Taking a really long view, honor thy father and thy mother means protect the environment of the algae that our own days may be long upon Earth.

2. Tracheophytes

Tracheophytes are the higher plants. They have a vascular system (a duct system that carries sap). Some divide this category into the Pteridophytes (plants that have a root, stem, and leaves), and the Spermatophytes (plants that also have seeds).

Ferns, 100 million years after the Cambrian, have leaves but reproduce by spores in a case. Gymnosperms with exposed seeds (conifers) appear after 50 million years more. After another 110 million years, we have angiosperms (flowering plants) with seeds in an ovary.

> **Comment:** Despite this progression of reproductive methods, no one has proposed an "evolutionary tree" for plants. This also challenges the Biblical version that plants were made before the animals.

Evolution

Evolution holds that, however life began on Earth, the first forms were single-cell organisms without nuclei that (somehow) gradually but inevitably developed into multicelled organisms with nuclei and so on to humans. This is monophyletic development: every organism is descended from a single ultimate ancestor.

> **Comment:** When I first heard of single-celled life, I thought of a tiny closed bag of some kind of liquid, but did not even think of what might be in that liquid.

A. Arguments for Evolution

1. Life from the lifeless

Evolution extends to the belief that the single cell organisms developed (somehow) from lifeless chemicals such as carbon dioxide, ammonia, methane, and water. The early Earth could have had a reducing atmosphere of hydrogen, methane, perhaps ammonia, and very little free oxygen. If so, the metabolic processes of the first life forms would have been radically different from those suited for an oxidizing atmosphere.

Perhaps there was a gradual change from reducing to oxidizing. Somehow the early life forms (pre-bacteria) responded by developing a membrane to protect their metabolic processes from too much oxygen. Maybe the oxygen oxidized the proteins on the surface of the prebacteria. But how did the surface become porous and how did the mechanisms develop to control the two-way flow of nutrients and waste products?

This would require some **pre-existing** self-organizing capability in the complex chemicals of the prebacteria. Science has made a great variety of chemicals, many that are unknown in nature. We have seen absolutely no evidence of such a capability in any of them. But Science refuses to apply its Doctrine of Uniformitarianism and conclude there never was such a capability.

The argument for Evolution seems based on the observation that the vast number of life forms we have found show a gradation of complexity in body structure and functions. The fundamental question then is did these arise naturally from simpler forms or did each one represent a separate act of Creation?

For all Metazoa, the fertilized egg apparently goes through some of the identical early stages. There is a single cell, then two, soon a cluster called a morula. This (somehow) develops into a hollow sphere called a blastula. Then (somehow) it folds inward to form a cup-shaped double-walled gastrula.

> **Question:** If we were all created separately, why is so much the same?
> **Answer:** Perhaps God created a basic cell that has almost infinite variability.

There is another set of organisms - viruses - that are made of organic material so they are not minerals. They are intermediate between lifeless chemicals and living entities. The tobacco mosaic virus has a crystal form. Viruses are parasites that grow on or in other life forms, implying they came after those forms.

- Science hasn't a clue as to how they arose. No common ancestor or genetic similarity is evident.

- The molecular weight of virus DNA may be as high as 2 trillion. Even at 1 million, since carbon, nitrogen, and oxygen have atomic weights of 12, 14, and 16 respectively, the molecule would contain roughly 70,000 atoms. We know of only one natural process that can make it. It is called life.

Evolutionists must map out a very long path from lifeless chemicals to virus DNA, let alone a single-celled animal.

2. Tree of Life - Science version

Many strongly reject the hypothesis that humans are related to the apes through a common ancient ancestor. However, except for the sponges and jellyfish, Science proposes that all animals are in the main line of Evolution which proceeds from protozoa (single-celled animals) to flatworms before branching off to vertebrates and other life forms. But consider:

Job 17:14 "I have said to corruption (decay), Thou art my father: to the worm, Thou art my mother, and my sister."

Mark 9:43 "And if thy hand offend thee, cut it off: it is better for thee to enter into life maimed, than having two hands to go into hell, into the fire that never shall be quenched: Where their worm dieth not (Jesus is speaking) . . . ."

We now attribute the decay in the Job quotation to microorganisms. The underlined words, repeated in **9:46, 48**, are strange. The **NIV** "explains" that worms were always found in decaying refuse, implying that worms "dieth not".

Question: But what did Jesus mean by "their" worm? Did He thus indicate a knowledge of some vestige in us of our primeval beginnings?

The Bible mentions a Tree of Life. Science has its own. **EST**, Animal evolution, has a chart of relationships but does label it as hypothetical. Lines are drawn connecting the various life forms shown, but they all lead back to a single cell at the bottom. "Somehow" the plants split off at the very beginning and never interfered with animal development thereafter. The sponges also split off by themselves.

The main line proceeds from protozoa to flatworms and then splits into two lines. One of these leads to fish and their direct descendants, the amphibians and the reptiles. Birds split off from reptiles and we get to the mammals, which are shown as direct descendants of reptiles, but no relationships are shown among them.

The same article has a second diagram which also starts with protozoa and shows the Chordata as descended from the protozoa. Neither diagram shows common ancestors at each branch point despite the vocal element of Evolutionists who insist they were there. Nor are any intermediates as postulated by Evolution shown.

3. Adaptation and mutation

Perhaps the branching in Evolution occurred at the DNA level through mutations. Radiation is known to cause alterations in our genes. We also know we are bombarded by many kinds of radiation.

Comment: Most mutations are harmful, even lethal. Be aware that in Evolution, if 99.9999% die, that is unimportant; the 0.0001% are a new species.

Resistant forms of both viruses and bacteria have appeared. Here is adaptation, not Evolution. The resistant form has acquired a defense against our medicines, but still causes the same disease. Some hasten to call the adaptations new viruses. What are the criteria for identifying a life form as a new species if it reproduces asexually?

Comment: Religion shows a similar mutation and adaptation capability: things people believed in and worshipped in the past no longer have any followers. The present Christianity includes many ideas found in other religions. We should not be disturbed by this. It reflects the gaining of understanding by humans of the nature of God, whether by inspiration, instruction or insight.

B. Arguments against Evolution

1. General

- No intermediates have ever been found anywhere in any strata.

  Comment: If the myriad of differences between species are due to gradual changes, the number of fossils of intermediates should vastly exceed the billions of actual fossils found. Failure to find intermediates does not of course disprove Evolution. It only says Science has been unable to find objective evidence for its

345

hypothesis. However, no other hypothesis has survived for so long without supporting evidence.

• Darwin's own writings show he would have discarded his theory had he known even aggressive search for intermediates would not find any.

• The millions of Cambrian fossils are of all kinds of creatures, some simple but also many complex. <u>No</u> precursors. And not one multicellular fossil that might have served as a precursor has ever been found in Pre-Cambrian rocks. Neither of these observations is consistent with Evolution. The specific difficulties are 1) finding complex forms in the earliest layers and 2) finding them as early as simple forms.

• Creatures from millions of years ago exist today essentially unchanged. Stasis. This shouldn't be.

> **Comment:** Not a good argument. Score one for Evolution. A mutation is independent of its parent. The parent can continue to spawn unmutated copies of itself indefinitely. There is, however, a question as to what the mutation mates with. Perhaps whatever caused the mutation made several similar mutants. Or perhaps the mutation was passed on in dominant genes.

• An amazing statement made by an evolutionist Dr. G. G. Simpson is that 2/3 of evolution was over by the <u>beginning</u> of the Cambrian Period (**DE**). So contrary to the protestations of evolutionists that a lot can happen in 4.5 billion years, the later changes must be done in only 440 million years, i.e., 10% of the life span of our Earth.

• **PCAI**, I, p 414, draws two amazing conclusions.

> o The Division Bilateria in the Superdivision Enterozoa is polyphyletic, meaning that its members arose by <u>several</u> different paths from Radialia (the coelenterates). This disagrees with **EST**, which says we came from flatworms.

> **Comment:** Evolution has to explain how, not one, but several independent paths developed.

> o In Bilateria, the fact that Brachiopoda are intermediate between Protostomia and Deuterostomia does not mean the latter came from Protostomia.

Comment: This logic says that even if we found intermediates among fossils, this could not be used as proof of direct descent!

2. Protozoa

Even the simplest protozoa are incredibly complex. Within a membrane, a bacterium contains cytoplasm with many kinds of enzymes, ribosomes to make proteins, mitochondria, lysosomes, Golgi bodies, reticulae, and a nucleoid containing genetic material. It has a flagellum with a helical twist and is driven by a molecular motor. Yet such life forms are supposed to be the earliest animals on Earth.

Question: How could they have arisen from inorganic substances? The bacterium simply could not exist unless it had all these parts in good working order. How could all these components arise simultaneously? Unless they were created.

To most, "enzyme" is just a word. Without enzymes, a bacterium could not make the hundreds of chemicals for its life processes. The incredible fact is that the bacterium also makes its own enzymes. Some are made for its ordinary needs of breaking down food particles and reassembling the parts into proteins it needs. Some are made to inactivate antibiotics. To make a particular compound may require several enzymes. Most enzymes are unstable outside of the life forms in which they were generated.

Comment: Contemplate all the controls that are needed. Something has to tell the bacterium to start making a particular enzyme or compound. It has to get the raw materials together, start the production mechanism, get the product to where it is needed, dispose of byproducts, check when it has made enough, and turn off the production mechanism.

Comment: I'm not sure there are any byproducts. Even more incredible.

The foregoing information seems to totally destroy the case of the Evolutionists. There is no way to get Evolution started because there is no way to bridge the enormous gap between lifeless inorganic chemicals and the simplest living cells. We are asked to abandon our rationality and believe that stirring up a soup of methane, ammonia, water, and even amino acids will produce a living cell. The simplest animal cell would be either an amoeba with a nucleus and no cell wall or a bacterium with a cell wall and no nucleus.

3.   Development of an eye

Evolutionists have been unable to propose a detailed mechanism for the evolution of body parts. It is very difficult to believe that the diverse components of, say, an eye evolved step by step. What is the survival value of a lens without a retina or an optic nerve? Try explaining how any of these components developed. Our DNA now tells our cells how much to grow. But in the beginning there was no blueprint. Unless you believe in Creation. And how would the changes get encoded in the DNA?

> **Comment:** The first step might have been a differentiation in which some cell or group of cells on the surface of an early creature became light-sensitive. But how? Could <u>any</u> cell in the body have become sensitive? Well, one of the characteristics of a life form is irritability, meaning it is sensitive to stimulation. So presumably any cell could have responded to light.

But what does the organism do with such a stimulus? Plant leaves are able to turn toward light. How? If there is a mechanism, we don't need to postulate a consciousness in a plant. But there must be a control mechanism to send instructions to certain plant fibers to tighten and perhaps others to loosen. Or does this mean that somehow there is communication between cells?

And there has to be a feedback mechanism so the leaf knows when to stop turning. It is interesting that the leaf does not oscillate to a final position. It seems like a response on the basis of diminishing returns. So no more than a non-conscious response seems enough to explain this behavior.

And yet: how is this built into a cell or group of cells? How did it arise? Or was it always there going back to the single-celled bacterium? If so, this means the chemicals in the membrane enclosing the workings of a bacterium had to be light-sensitive. Well, all green plants have complex molecules like chlorophyll, and in the beginning the blue-green bacteria contained chlorophyll.

4.   Chlorophyll

Chlorophyll in plants is analogous to hemoglobin in animals. There are several chlorphylls, one of which is chlorophyll a: $C_{55}H_{72}MgN_4O_5$.

> **Description:** A pyrrole ring is made of four carbon atoms and one nitrogen atom with one hydrogen at each of these five atoms: $(CH)_4NH$.

Four such rings minus all their hydrogens and joined by single carbon atoms form a porphyrin ring:

-(CNC_3)-C-(CNC_3)-C-(CNC_3)-C-(CNC_3)-C-

The terminal atoms join to close the ring. The nitrogen atoms point toward the center of the ring where a metal atom (magnesium for plants) is in an electron-sharing configuration. Some idea of the complexity of chlorophyll a is obtained by writing the formula as follows:

$$Mg[(CNC_3)C]_4[(CH_2)_2COOC_{20}H_{39}][OCCHCOOCH_3]-$$

$$-[C_2H_5][CH_3]_4[CHCH_2][H]_5$$

The groups in brackets are attached to the central ring, given by $[(CNC_3)C]_4$, in various places. To make chlorophyll a, 137 atoms have to get together in the right configuration by chance collisions.

Question: Imagine what it takes to make such complex compounds from scratch. As far as I know, it has never been synthesized at <u>ambient</u> conditions. Monovinyl chlorophyll a has been made in the laboratory. How could the various atoms have been persuaded to join under the conditions in nature? Even if it was made under early Earth conditions, it was useless if no organism were present to use the photosynthetic energy it makes.

5.   Cell division

We can pose another challenge to Evolution. You may have a general notion that single-celled animals like bacteria reproduce by mitosis. At a certain time in the life of a bacterium, its single chromosome will divide in half. The two halves will withdraw to opposite sides of the cell. The cell will pinch down between them and finally split into two separate cells.

Comment: Come now. You have to be astounded that a mechanism in these tiny cells compels the chromosomes not only to split at a certain age but to move in an orderly fashion to opposite sides of the cell instead of just milling around. What is controlling the exquisite timing and movement? And how did it get started? Science again puts on the blinders and says, "It just does."

Once I saw a television show on how human cells divide. I watched a fertilized egg cell A_o divide into two A_1s, then into four A_2s, and so on. But

349

suddenly I understood what I was seeing. Somehow the cell had learned how to divide but stick together.

So we don't have to worry about difficult questions like whether reptiles gave rise to birds, or how did man arise from reptiles. There is a much simpler question to explore in the laboratories: what makes the A_1 cells stick together? And in the second step, each pair of A_2s sticks together but also to the A_1s they came from. Why? How did this capability develop? Why did it develop?

On further thought I realized I had misinterpreted what I was seeing. We go back to how single cell animals reproduce. After the chromosome halves have moved to opposite sides of the cell, a septum forms between them before the cell starts to divide by pinching down its middle. So what has happened in the multicellular animals is that the last act, the pinching down, doesn't occur. But then we have the question of how this deletion came about.

Summary - Whereas:

1. Our Earth appears to be at least millions of years older than 10,000 years.

   • Dates for objects older than ten thousand years can be estimated only by radioisotopes. Some igneous rocks have been dated by radiometry to 3800 million years. We really need to check the uncertainty in these dates.

   • Sedimentary rocks do not contain radioisotopes. Yet these are the rocks that usually contain fossils. There is a serious question as to how the ages of the rocks and fossils were established.

2. In many places the structure of our Earth shows layers, or strata, that make up the geologic column.

   • The strata are believed to have been formed by recurrent flooding due to changing ocean levels that are attributed to similar changes in sea bed levels.

   • There is evidence of at least 16 megafloods in North America (**TOG**).

3. In the strata are found fossils, which are evidences of long-dead life forms.

• Known fossils are no older than about 650 million years. Most are of marine creatures, which means the strata in which they were found were once under water. The fossils survived because they were buried soon after death and/or mineralized.

• About 70% of known animal Phyla are represented in the earliest strata in which fossils are found - the Cambrian. This is a radical contradiction of Evolution in that complex life forms are found as early as simple ones.

• Not one precursor or intermediate life form has ever been found anywhere.

• Speculations as to initial life forms include only bacteria and algae, possibly back to 3500 million years ago. Algae require certain vitamins that are produced by bacteria, so unless this need was not present in the earliest algae, bacteria must have been the first life forms.

• Bacteria also need vitamins, which some can synthesize. Others have lost this ability, which has Evolution proceeding in the wrong direction.

• Bacteria can make enzymes and coenzymes they need.

• Chlorophyll is a complex molecule without which no life could have arisen or continue to exist on Earth. An enormous faith is required to believe it could have come about by random reactions among lifeless inorganic chemicals.

• Spontaneous generation of bacteria was thoroughly disproven in the 19th century. Science still clings to the belief that it could have happened in the remote past to produce ancient bacteria. This violates its absolutist Creed of Uniformitarianism.

4. Science has no consensus on how to classify life forms.

• Four different systems have been devised. The evolutionary system attempts to show ancestor- descendant relationships. This has proven to be impossible, so the cladistic looks at branching.

• Diagrams on evolutionary developments do not show any intermediates or common ancestors.

- There are obvious similarities in body structure for various Animalia, but no mechanism has ever been proposed that might account for the changes in body structure that Science wishes to attribute to a mysterious impersonal Evolution.

- If Evolution is true, it had to have accomplished 60% of its work <u>before</u> the earliest fossils, the rest within 440 million years, not 4.5 billion.

- Viruses cannot be classified in the same way as plants and animals. Science has no idea how they arose. There is no common ancestor nor any evident genetic relationship.

- The molecular weight of viruses may go as high as 2 trillion. Even one with a molecular weight of only one million would contain roughly 70,000 atoms. The only known process for assembling it is life.

It therefore appears unavoidable to conclude that:

- However life first appeared on Earth, it is not credible that the large number of complex chemical substances required to initiate and sustain life in its simplest forms came about simultaneously by strictly random processes.

- It is not reasonable to believe that random processes could have provided a series of initiating causes that would consistently lead to the development of life forms with ever more ordered and complex organization and performance capabilities. All natural processes go towards increasing randomness and disorder.

- The stagewise development of the Subphylum Vertebrata in particular indicates a corresponding series of initiating causes was involved.

- Even if an "evolutionary" mechanism is conceived, it would not in itself explain how it came about any more than explaining how a mechanical clock or a gasoline engine works explains how they came about.

- Therefore, the initiating causes constitute a series of unidirectional nonrandom events.

At issue then is the choice of believing that an inorganic Universe can spontaneously generate such a series of nonrandom events or that the intervention of an outside Intelligence is required.

There are, however, many questions that Religion needs to address to reconcile its teachings with what has been found by observation of our Universe. E.g.:

- The Biblical Flood may be better explained by subsidence of the Earth's crust than by rainfall. Consider **Genesis 7:11** ". . .the same day were all the fountains of the deep broken up. . . ."

  Comment: This implies humans lived through the breaking up of Pangaea.

- The coelenterates (medusae, jellyfish, corals) were possibly the first predators. They had tentacles to seize prey. They existed before humans. Creationists need to consider this in connection with the doctrine of original sin introducing death into the world.

Another approach

We have gone far enough down this lengthy and murky path. Let us try another tack. Although our bodies are mostly water, our tissues are made of various compounds of carbon, hydrogen, oxygen and nitrogen. There is some sulfur and phosphorus, with little bits of sodium, potassium, calcium, magnesium, and iron thrown in. So let's ask some questions.

1. HOW can and WHY should a pile of such atoms:

- be conscious, be able to think about itself;
- want to think about itself;
- be curious, pose questions, and crave answers;
- have a drive to seek, to explore;
- crave certainty, be intolerant of uncertainty;

- want to stay alive;
- crave stimulation;
- want to be secure, yet take great risks;
- want to build, yet want to destroy;
- want variety yet constancy;
- want to be free;

- want to be different, yet suffer from loneliness;
- want to mate;

- want to compete and win, yet want to cooperate;
- want to protect, yet want to kill;
- want approval, and shun criticism;
- show empathy/rejection;
- want to be seen as good and right;
- have an <u>immediate</u> sense of having done wrong, even as a child;

- love/hate, trust/fear;
- laugh, cry;
- weep, rejoice;
- feel fulfillment/disappointment, pride/shame, innocence/guilt;
- feel gratitude;
- feel emotional and spiritual pain;
- feel elation/despair;
- feel boredom/excitement;
- feel generous/jealous?

2. What is the source of:

- optimism and pessimism;
- faith, hope and charity;
- intuition?

3. Who or what is "I"?

The scientist who avoids these questions, or denies their existence, has abandoned his humanness as well as his intellectual integrity and become a robot. We need to tie down the loose ends of explanations. God might be conducting a colossal experiment in conferring consciousness and humanity on matter.

Consider an analogy: a computer. Let's think of a sophisticated model, one that is mobile. Let's also suppose that it has a battery as a source of power.

- How does it get turned on? Well, its program could flip a switch.

- When would it flip the switch? Its program could specify a time of day, or perhaps it could be activated by a photocell reacting to daylight. It could even be made to turn on at a random interval of time after some preselected reference time.

- But is there any criterion corresponding to human choice? The computer does not have to entertain itself, seek stimulation or pass time. But its program might require it to keep itself in good repair.

- Its program could initiate a search for a station offering useful information on programming.

- The computer could be designed with arms so that it could insert a plug into an electric receptacle. When? In response to some criterion, like battery voltage or hours in service.

All of this is in the material world. When we concentrate on the material, we lose forever those precious moments when we could experience the mental and spiritual capacities of our being. I believe there is something more than the material world. Even if it cannot be measured. We are pieces in a cosmic jigsaw puzzle. Let's accept insight from whatever source and cooperate in discovering what the picture is about.

Unexplained Phenomena

There are also phenomena which Science cannot explain and chooses to ignore. These cover psychic phenomena, miraculous healings, and, yes, even UFOs. There are thousands of such experiences, but no "scientific" explanation. Examination of these phenomena is outside the scope of this book. I have never seen a UFO, a ghost, a demon, or a miracle, although the fact of our existence can be seen as a miracle on the basis of sheer improbability.

Definition: A miracle is an event in the physical world that surpasses all known human or natural powers and is ascribed to a divine or supernatural cause (**RHCD**).

Thus, if someone grew a new foot to replace an amputated one while we watched, we would call that a miracle. I have never heard of such an event. But I have heard of people recovering after medical practitioners have pronounced the case for their recovery hopeless and the patient clinically dead. This includes cancer, paralysis and coma.

These reports are noted to encourage active exploration of the possibility that there is much more to our Universe than just the three physical dimensions and the one time dimension we readily perceive, and that Science condescends to consider.

So in the gamut of knowledge, we have hunches, ideas, concepts, possibilities, probabilities, demonstrated facts, and ultimate truths. What shall we do now? For one thing, I will update Omar Khayyám:

> Myself when old still eagerly sought
>> The Doctors of Science and what they taught
> But only did find that most were blind
>> To the wonder of what God hath wrought.

For another, let us return to identifying attributes and principles that are important to our survival, and then to forming our own conclusions as to our question.

Chapter 16. Eclectic Pragmatism (EP)

Unfortunately, there still seems to be no one idea, no single system that we are <u>all</u> willing to accept as a common guide for use in our daily lives. We find that everything requires interpretation. There is always an exception, an extenuating circumstance, or an elusive aspect that has not been explained. Even the infallible Bible is being offered in ever more current and correct "clarifications".

You may also have serious objections to any kind of "pragmatism" - hasn't it been tried before? Yes, and its teachings linger on. Two names - James and Dewey - are especially associated with it. We need to review what it meant to them and to make clear how EP differs from it. This will not be easy. Translating what philosophers say to lay bare their meanings is usually a tedious and dreary task. Let us therefore draw upon the critiques of others to aid us in this task.

William James (1842-1910)

<u>A. James' ideas</u>

1. Basic doctrines

James called his system of doctrines "radical empiricism". A doctrine is any opinion or dogma held as true. This says nothing about proof. Therefore, read carefully, for he was viewed as the <u>leader</u> of American philosophy in his time.

"Empiricism" is the doctrine that <u>all</u> knowledge is derived from sense experience. A competing doctrine is "rationalism", which appears in two forms. In philosophy, reason alone, <u>independent</u> of experience, is <u>a</u> source of knowledge. In theology, human reason, without divine revelation, is an adequate or sole guide to all attainable <u>religious</u> truth (**RHCD**).

> **Comment:** These assert that humans need only use their a) experiences or b) minds to discover all truths, some truths, or religious truths. EP holds that reason alone cannot escape errors. Whether revelation is required will not be discussed here.

A "radical" change is a thoroughgoing, fundamental, or extreme change. A "Radical" favors drastic political, economic, and social reforms by <u>direct</u> and <u>uncompromising</u> methods (**RHCD**).

357

Comment: Any thoughtful person should now flee from "radical empiricism" with a shudder over the implied ruthlessness. But "empiricism" alone is suspect in asserting knowledge cannot be obtained from sources other than sense experience, such as inference, intuition and revelation. It thereby a) denigrates human capabilities and b) repudiates Religion.

But let's look a little into his pronouncements. James <u>asserted</u> that there is only one primal stuff or material composing everything in the world, and that is "pure experience". It is the immediate flux of life which furnishes the material for our later reflection. It is neither mind nor matter (**HWP**). He also said that experience is not made of any general stuff (**PITC**).

Critique: Experience cannot exist without life. By common sense, some events in our lives are not noticed, hence not experienced; also, our lives are influenced by memories of former events. So experience does not extend equally in space and time with life (**HWP**). We can salvage something if we say our experiences with each other and the Universe give us the empirical data from which we infer rules of behavior to maximize our survival probability.

Pragmatism is a method for indicating ways to change existing realities. It turns away from abstractions, fixed principles, closed systems and pretended absolutes toward facts, action, and power (**PITC**).

Comment: EP focusses on preserving elements that have contributed to the advance of civilization while providing for orderly adaptation to changing material and social circumstances.

Pragmatism turns away from first things and principles toward last things, facts, and consequences (**PITC**).

Comment: Isn't this the same as the end justifies the means? This concept of pragmatism means our rules won't hold tomorrow, so don't make contracts.

2. Religion

A theory of the universe would have good consequences only if it included at least moral responsibility and freedom of action. He did not deny there was a world outside of experience causing experiences but held that we could not experience that world itself and so could know nothing about it.

Our world is a world of human experiences since it is through them that we interpret the world; it includes science and human values (**BTGP**).

> **Comment:** What is "moral responsibility" and on what is it based? What are "good" consequences? We can experience electricity and while we know nothing about its nature, we do know enough about it to use its properties. The last statement is false: both Science and Religion agree there once was a Universe without humans. <u>Our</u> experiences thus cannot explain pre-existing matter and energy. It is true if he means we cannot <u>affect</u> our material Universe or any world outside it but can have some effect on our human experiences.

An idea is true if believing it results in some gain to us. The task of philosophy is to find the effect on us if some hypothesis about the world is true or false. In religion, we cannot reject any hypothesis about God if useful consequences follow from such belief (**HWP**).

> **Critique:** We can thus concentrate on the belief and forget about the object of belief (God). James' doctrine thus depends on fallacies that arise from ignoring extra-human facts. **HWP** concludes with: " . . . this is a form of the subjectivistic madness characteristic of most modern philosophy."

> **Question:** Since believing in God results in a gain (hope, relief from despair, peace of mind, etc.), why didn't he conclude that God exists? He ignores the fact that a practical consequence of the unshakable belief in the divinity of Christ was the victory of Christianity over the Roman Empire and its gods. By his own thesis, Christianity is true. However, Marxism overthrew the Romanoffs in Russia. So Marxism is also true. If pragmatism is to be viable, it must be concerned with more than just gain. The test of a belief must consider <u>all</u> of its long-term consequences.

The existence of God is not to be decided by reason, but only by a faith that makes a practical difference to man (**HET**).

> **Comment:** Such as reward or punishment. This bold assertion would surely be seen by God, if He is real, as arrogantly self-centered.

Belief in God is necessary for the satisfaction of our nature. (Somehow) we have a will to believe in God. This God is part of the universe, and is working with man in the realization of man's ideals. Belief in an immortal soul is useful in our moral lives, but he did not include it in his philosophy (**BTGP**).

> **Comment:** James didn't explain the origin of our will to believe. His notion of God sounds partly like an intelligent personal Being. But if God is only a part of the Universe, where did the rest of it come from? However, if anything is going to be realized, experience shows it can only be God's ideals: all of man's eventually disappears. And how can an idea be useful if it is not included in one's philosophy?

He affirmed the values of liberal Christianity, and was opposed to all forms of religious orthodoxy. Man was to be guided by faith (!) and sustained by hope (**HET**).

> **Comment:** Faith in what? Hope of what? Without the theology of orthodoxy, there is nothing to have faith in or hope for, and no reason to think these words are anything but empty air. We can only pray - yes, pray - that our rulers will not change their minds as to what rules shall hold tomorrow. "Liberal" here is exposed as merely denial of supernaturalism, and hence denotes at most a secular religion.

He opposed absolute idealism and absolute values. Values arise from human experience and depend on the way we feel and act. They represent smugness, complacency, and dogmatism (**HET**).

> **Comment:** What values is he talking about? Christianity has a prohibition on murder. It has nothing to do with how the victim feels, since he is dead. Are the values of his liberal Christianity also smug and complacent? If they arise only from experience, clearly there are no common values because my experiences are different from yours. Though we rarely attain in our lives the absolute in values such as truth, goodness, compassion, etc., having such concepts gives us direction for our actions. Values became absolute at least because they were tested by human experience. They can be changed if further experience results in a clearer understanding. This is exactly parallel to Science holding that the velocity of light is constant (with exceptions) until further data show this absolute is untenable. Or to Religion, where God has changed from a fearsome vengeful deity to a God of love.

B. Assessment

What picture of James do we get? Someone chafing at the restrictions and inequities of society, objecting to the rigidity of theology, criticizing preceding philosophers, but holding on to an undefined vision of a modified relation between man, nature and God.

EP certainly differs from James by holding that truth is not measured by gain. What is the practical consequence (gain) from knowing it is true that George Washington was the first president of the United States? We believe we cannot see the dark side of the Moon from Earth. Does the lack of gain mean it is false? Truth is also not necessarily "good". It is true that a sharp object can cut, which leads to tools and weapons. These can be used for construction and defense, <u>or</u> destruction and offense. In EP, the focus is on what accomplishes some purpose (hence, pragmatic). The selection of the purpose depends on criteria involving more than EP alone.

John Dewey (1859-1952)

Dewey was a more formidable thinker and a prolific writer. **BTGP** labels him as the leader of present-day (c1940) pragmatism, focussing on his belief that it makes no difference to us if there is or is not another world besides the material world of our senses. Dewey wants to explain how our experiences come about and how they influence other experiences.

I started reading **IMW**, an 825-page collection of his writings, some fifty years ago. A peculiarity is that it adds a 239-page introduction by someone named Joseph Ratner, identified only as an editor. This seemed to be an extraordinary imposition on the reader, so I ignored it completely because I wanted to know what <u>Dewey</u> said. However, I found his style was abstract and obscure, words were given unusual meanings, its underlying message was unclear, and I had no immediate need of any of its conclusions. I soon set it aside. With the hindsight of 100 years of bloody history there is a powerful incentive to understand what he was urging upon us as his kind of pragmatism.

<u>A. Dewey's ideas</u>

    1.   Basic doctrines

Dewey first called his system "instrumentalism" and later "experimentalism". **RHCD** says instrumentalism is the variety of pragmatism developed by Dewey that says an idea is true if it successfully solves a problem and the value of ideas is determined by their function in human experience.

> **Comment:** This is too broad. Consider the value of theft as a useful and successful (sometimes) solution to your need for money.

Dewey asserts there are no Absolute Truths. Except his own. Here are a few samples:

1) The cosmos is unimportant to education.
2) Man can be changed.
3) No one has "true" individuality except as a member of a group.
4) Morality comes from spontaneity, not from the dictates of the past.

He approved of Charles Peirce's definition of truth as the opinion ultimately agreed to by the investigators.

> **Comment:** Consensus is thus an approximation to the truth. This procedure was used in agreeing on the velocity of light, showing that even in Science truth is a matter to be decided by majority vote. Remember: the majority vote before Copernicus was that the Sun went around the Earth.

> A more serious consideration is that if a consensus of differing opinions is the best that Science can achieve, can we expect anything more in the area of values? But surely this is better than a consensus in which no one investigates.

Dewey judges a belief by its effects, and so talks of "warranted assertability" rather than "truth".

> **Critique:** Effects are future events. If we alter the future, we can alter truth, i.e., what we say about the past (**HWP**).

> **Comment:** Dewey's argument thus sounds like the foundation for revisionism. Also, we must know what truth is today so we can make decisions today.

To Dewey, a belief may be good at one time and bad at another. It is good if it causes the organism (that's us) to engage in activities that have satisfactory consequences; otherwise it is bad.

> **Comment:** Perhaps this spawned relativism and **SE** ("Situation Ethics"). Shouldn't Dewey therefore have believed in the resurrection of Christ since many good things resulted from this belief?

He did not seek "absolute truth" or "absolute falsity". There is a process he called "inquiry", defined as:

> "The controlled. . .transformation of an indeterminate situation into one that is so determinate in its constituent distinctions and relations as to

convert the elements of the original situation into a unified whole. It is a mutual adjustment between an organism and its environment."

Critique: This definition is at least inadequate: it applies to a bricklayer making a wall (a unified whole) out of a pile of bricks (**HWP**).

Comment: "Transform" means to change <u>completely</u> (my emphasis) in form, appearance, structure, condition, nature, or character. Dewey surely was not content with changing only appearance, form or condition. Regarding "adjustment", natural situations make absolutely no adjustment to our needs, so he must be talking only about human situations. Note the insinuation of the element of control. Who is doing the controlling? In what way was the original situation indeterminate or not unified?

Comment: To call a transformation an inquiry is to obscure what is intended by inquiry. This stirs recollections of Orwell's "1984". Let us be aware of the Newspeak potentials of "adjustment".

Comment: "Inquiry" applies to <u>all</u> human relations because they are all indeterminate to some degree. Paraphrased, it is: the controlled change from an existing relation among several independent parts to a single unit in which the parts have lost their independence. It is a formula for replacing everything in an existing society to form a monolithic society. "Monolithic" means total uniformity and intractability. This is totalitarianism.

Following Hegel, "inquiry" also applies to whatever "unified whole" is constructed. So there is no stopping the process at any particular "whole", although totalitarianism has proven to be a difficult "whole" to modify.

2.   Religion

We should not interpret some of our experiences in terms of God because we cannot prove very much in those interpretations (**BTGP**).

Question: Why is Dewey exempt from his own objection? He asserts interpretations based on unproven, and probably insupportable, assertions.

He ridicules religion as meaningless since its moral motivations range from temple prostitution to vestal virgins, and there have been so many incompatible ways in which the unseen power has been worshipped.

Comment: His arguments deal merely with the inadequacy of human efforts to understand and relate to God, not with the existence of God. They change because in their "inquiries", they have found many inadequate answers. He refuses to declare explicitly that he believes there is no God.

He asks why should we assume that no further change in our concept of the unseen power will occur? The logic involved in getting rid of inconvenient aspects of past religions compels us to inquire how much now accepted are survivals from outgrown cultures.

Comment: He ignores the fact that, despite the number of forms, all have conceived of God as a supernatural Being. That anything survives at all is exactly parallel to Science where one misconception after another is exposed and discarded.

Question: What are these inconvenient aspects (my emphasis)? What is this "logic" of riddance?

Supernaturalism gets in the way of an effective realization of the (broad) implications of natural human relations, and of changing them radically.

Comment: Dewey talks in abstractions and insinuations and spews out word pictures but never comes right out to illustrate what he means. What does supernaturalism get in the way of? What are these "natural human relations"? What are their implications? What are these radical changes?

Religion tells us it is "natural" for humans to lie, steal, fornicate, commit adultery, and murder. These are all forms of taking without permission what doesn't belong to us. We have learned that we must restrain ourselves either 1) by our morals or our will or 2) by the State so that we may enjoy less stressful relations with one another because they will become freer, less guarded. Does Dewey propose we abandon all restraints because it is unnatural to refrain from taking?

We should start afresh by asking what would be the idea of the unseen, the manner of its control over us and how reverence and obedience would be shown if whatever is basically religious in experience were allowed to express itself free from all historic encumbrances.

Comment: In short, let's start a new religion. Let us work out a new Fifteen Commandments, or whatever. Humanity has been down this path. Many times. What we have today is the result of untold hardship and struggle to understand our God. We don't need to start afresh. If there is something wrong in Religion, perhaps it is because some critical element has been corrupted or lost, or some alien element has crept in that must be excised. Christ was very clear about the essence of the law:

> **Matthew 22:37** ". . .Thou shalt love the Lord thy God with all thy heart, and with all thy soul, and with all thy mind."

> **Matthew 23:39** ". . .Thou shalt love thy neighbor as thyself."

Before we discard all the results of our previous experiments, let us understand what lies at the end of Dewey's inquiries and experiments.

Aggressive atheism has something in common with traditional supernaturalism. It is so negative that it fails to give positive direction to thought. But more importantly, both are exclusively preoccupied with man in isolation. The only thing of ultimate importance is the drama of sin and redemption played out in the lonely soul of man. Militant atheism also lacks natural piety. Its attitude is the universe is indifferent and hostile and we must defy it.

> **Comment:** More assertions, without one shred of the proof he demands of Religion. His hostility to traditional Religion is apparent. Supernaturalism has nothing in common with atheism and gives very positive direction to thought, as in **Matthew**, above.

A religious attitude needs the sense of a connection of man with the developing world through dependence and support. "God" is a working union of the ideal and the actual. The meaning of God involves selecting the factors that generate and support our idea of good as an end to be striven for.

> **Comment**: What is "our idea of good"? It sounds like Dewey believes in an Absolute Good. Unless his God changes with experience. What are we to be dependent upon and supported by? In any case, this is a can of words with a meaning so fuzzy that even supernaturalism can be included as a support factor.

A humanistic religion that excludes our relation to nature is pale and thin, and is presumptuous when it takes humanity as an object of worship. The only alternative is supernaturalism.

> **Comment:** This is a complex statement. It says our (sinful, animal) natures must be included in religious practices. But then it concedes we

365

must not worship ourselves. On what basis is a humanistic religion to be built? If impossible to construct, will Dewey then accept God?

He seems unable to accept the drive of humans to refill any void created by eliminating supernatural aspects of existing religions, be it from curiosity, boredom, insecurity, or faith. The continuation of belief in the supernatural does not mean people are ignorant savages, but rather they comprehend that humans need to relate to an outside Power to cope with the Universe they find themselves in.

Near the end of **IMW** Dewey finally makes a kind of a definition of "religious": any activity pursued in behalf of an ideal end against obstacles and in spite of threats of personal loss because of conviction its general and enduring value is religious in quality. But he insists that the identification (of his idea) of religious values with the creeds and cults of religions must be dissolved.

Comment: Commitment to <u>any</u> ideal end is not necessarily religious. Reforestation is an ideal end but is often a matter of self-interest and is thus not religious. Dewey's zeal for "experimentalism" could make it a religion, although a limited one. A definition of religion has to address the nature of our relation to the Universe and to each other. After reading several of his writings, he still hasn't told me what <u>his</u> religious values are.

The value of religion was not its theology but its capacity to transform human behavior to the more humane, more compassionate, and more altruistic. Moralists, theologians and teachers of the past overlooked the instrumental values that would address questions like:

- How could the good life be achieved?
- What means and processes could be used to attain ideal values?
- How could theory and practice meet in man's moral life?

Comment: Dewey appoints himself as judge in deciding what the role of theology in Religion is. He fails to understand that the theology is what makes possible the transformation he desires. Without theology, there could be no basis for favoring one man's rules over another's.

3. Democracy

The foregoing sections were intended to examine Dewey's experimentalism for evidences of pragmatism. I can not know what impression of his ideas and his objectives you might have gotten from these. Some of his writings on democracy also appear inconsistent with them.

- No limited set of men is wise enough or good enough to rule others without their consent.

- All who are affected by social institutions must have a share in producing and managing them.

- Intelligence is sufficiently general so that each individual has something to contribute.

These strongly support the interests of the individual, and to my mind conflict sharply with Dewey's advocacy of group rather than individual effort. The effect on his philosophy of the interplay among these three ideas is very unclear.

B. Assessment

Dewey ignored the cosmos most of the time (**HWP**).

> **Comment:** Thus, what he proposes deals with only a tiny part of our Universe. Like trying to explain a dog by minutely examining a hair on its tail.

He was influenced strongly by Evolution: man can be changed since he is neither good nor evil.

> **Comment:** We now have proof that even under absolute totalitarian control the trait of self-centeredness was found to be inherent and readily expressed as self-interest and ultimately a ruthless selfishness. "Inherent" = "sinful nature". So much for the experiment of the workers' paradise.

> Man's behavior can be changed, but I do not believe his nature is changed thereby. Even Dewey agrees: in his "The Public and Its Problems" published in 1927 when he was 68 we find:

>> "The old Adam, the unregenerate element in human nature, persists. It shows itself wherever the method obtains of attaining results by use of force instead of by the method of communications and enlightenment (**IMW**)".

Critique: Russell thinks the belief in human power and the unwillingness to admit "stubborn facts" stemmed from the hope that man's condition could be improved. This arose from increased production by machines and the scientific manipulation of our environment. I believe the following captures the essence of his opinion on this point.

"(Humans) began to think of themselves almost as Gods. (This) is a great danger . . . a cosmic impiety. The concept of "truth" as something dependent on facts outside of human control has been one of the ways philosophy . . . (gave us) the necessary element of humility. When this check on pride is removed, (we move) toward. . . an intoxication of power. . . (which is) the greatest danger of our time. Any philosophy . . . contributing to it increases the danger of vast social disaster" (**HWP**).

Comment: For me, Dewey emerges as an implacable foe of traditional religion because of its insistence on the reality of the supernatural. I do not believe that the desirable course of action is to partition religion into a set of religious experiences, a set of religious beliefs, and a set of practices of religion and then totally discard the latter two. It sounds like wanting to have orgasms without the commitments of a relationship.

Christianity emerged as dominant over the gods of an all-powerful Roman state despite incredible tortures and fiendish deaths, and survived for two thousand years in the face of man's corruption both in and out of the church. This testifies that ordinary humans find something of great value in it that transcends Dewey's unwillingness to practice that openness of mind he preaches with regard to the beliefs of those who have through their sacrifices left us the priceless heritage of freedom, the very freedom that permits him to mouth his assertions.

Dewey preaches his pragmatism arrogantly as if there is no doubt it will work. He conveys an attitude of let's try any experiment in the short term in the matter of changing existing systems because he says there is something wrong with them.

One gets a disturbing impression partly because he does not define his terms clearly. He seems to talk on both sides of concepts. There is another disturbing note. As an apologist for Dewey, we find Ratner saying:

"One of the more bizarre of the absurdities now being given wide currency is that Communism is the source of Materialism . . . the illustrious fathers of Materialistic Philosophy are none other than Galileo, Descartes, and Newton (!) The differences between the

Newtonian and Marxian varieties of Materialism . . . unquestionably redound to the favor of of the latter. For Newtonian materialism, apart from its appalling intellectual poverty, is such a childishly dreary mechanical affair - an unimaginative push-and-pull business. But the Marxian Materialism goes along in ever more novel ways, developing itself and the universe in accordance with the magical antics of the Hegelian Idealistic dialectics secreted in its vitals . . . Every fair mind must admit that the Magic (confers) on the philosophy. . .the substance of organismic character. And almost any organismic philosophy, no matter how bad, is better than almost any mechanical philosophy, no matter how good (**IMW**, p 55)."

Comment: This is precisely the kind of sycophantic blithering that we must not identify with thinking. Perhaps the tens of millions who suffered expropriation, and were starved and murdered by their own Marxian governments were amused by the magical antics of Hegelian dialectics. Ratner noted that Dewey had read through page 56 and so we must conclude that he agreed with Ratner's writings at least to that point.

The close association between Ratner and Dewey is disturbing in view of what has happened in every Marxist country. We must be wary of Ratner's interpretations of Dewey and perhaps even of Dewey's own writings.

For the purposes of this book, I believe I can recapture the term "pragmatism" if I make clear its meaning by its characteristics. Eclectic Pragmatism aims at providing a meeting ground for those of many persuasions as the first order of business for resolving our many conflicts. Unless we can do this, there won't be any humans or world for us to debate with or about.

Eclectic pragmatism

One principle that emerges from human experience is the value of freedom to adapt to changing circumstances. We must not create a straightjacket around our bodies, a shell around our minds, or a cage around our spirits. We need rather a shield, a screen, and a sail.

Until it can be shown that our entire system is hopelessly inadequate for modern conditions, we should prefer evolution to revolution. It is curious that Dewey in a world of Evolutionists should propose revolution in human organizations, whereas Religion in a created world, which is a revolution, should propose gradualism, i.e., evolution. This is how we are proceeding in America. The alternative is the bloodshed we see in other countries who seem to have no alternative but revolution. And never imagine that if there is to be bloodshed, it will not be yours.

The pragmatism of EP is a much more direct and conservative form. It is not a religion. But whatever your belief system is, we need to hold some principles and truths in common. We must therefore search for certain system attributes and personal qualities that we judge to be of long-term usefulness in our interpersonal relationships. These need to be explicit in our belief systems. Perhaps we can also use some of the rules of Logic that Science has developed, and some of the elements of Faith that Religion has discerned.

My suggestions follow. These are collected under the name Eclectic Pragmatism, and are offered as a beginning. Let me expand first on the statistical aspect because we need to agree on the nature of the problem of constructing a system inclusive of all.

A. Statistical

1. Uniqueness

We hear talk about how unique humans are. To show the basis for this opinion, suppose there were three attributes that we each either had or did not have. There would be 2^3, or eight combinations of these attributes making eight kinds of people:

ABC   ABO   AOC   AOO
OBC   OBO   OOC   OOO

Now suppose there were 33 such attributes. This would accommodate a population of 2^{33}, or 8.6 billion unique individuals. Keep this in mind when you think of the problems in establishing relations between people. The uniqueness is even greater when we consider that there can be gradations in each attribute. If we used only three gradations - average, more than average, less than average - we would need only 11 attributes to characterize 8.6 billion people. And who would disagree with the observation that we each have far more than 33 attributes? External physical attributes alone easily add up to more than 33.

2. Dimensions in the material world

But this is not all the complexity we need to deal with. We live in a four-dimensional world of up and down, left and right, forwards and backwards, yesterday and tomorrow. The zero point in each dimension is where we individually see ourselves. These are the primary dimensions of our material world. A significant aspect of the statistical attribute is that it is continuous, ranging from nothing through a little bit, an average amount, a great deal, and to totality. Distance is truly continuous in ranging from minus infinity through zero to plus infinity. Others start at zero and go through increasingly positive values.

> **Comment:** Time is more of a descriptor than a dimension. We can combine the three spatial dimensions readily, but we cannot add three feet to twenty seconds. Time is also limited in that we cannot move backward through it, so we have access only to future values of time.

We can expand the application of this image. Our physical world includes the physical ways we spend our time and otherwise identify ourselves. We have different kinds of jobs, ways of getting around, housing, clothing, relationships, pastimes, etc. Each of these can be thought of as a linear scale, or axis, ranging from 0% to 100%. In effect, we live in a multidimensional polarized world. No matter how many dimensions one includes, and without considering how one actually represents these, we can imagine people distributed along each axis so as to represent how much of that particular characteristic each person has.

3. Other worlds

The previous pertains to the world of the body. I suggest we become conscious of living in at least two other worlds: the world of the mind, and the world of the spirit. What do I mean by living in the world of the mind? It means devoting some portion of one's waking moments to activities that are primarily mental: reading, thinking, planning, discussing, analyzing, theorizing, etc. We cannot forget the needs of the body, but they are taken care of to varying degrees according to the strength of the desire to be mental.

The world of the spirit is probably experienced mainly as a subset of the mental world insofar as the subject matter of one's thoughts is concerned with concepts of justice, relatedness, courage, etc. It is manifested by our feelings: elation/depression, trust/fear, delight/anger, etc. And at times we ponder on survival after bodily death, universal eternal judgment, etc.

The degree we are in each of these three worlds can range from 0 to 100%. Most of us are probably more than 50% in the world of the body, 0 to 5% in the world of the spirit, and the rest in the world of the mind. We can describe ourselves statistically by the applicable percentages.

It would seem unlikely for two people to occupy exactly the same point in this multidimensional Universe. Like Pauli's Exclusion Principles for quantum numbers.

What is the application? We need to keep such a picture in our minds that that is the way the world is and that is the world we must deal with. We are told that we are different, yet refuse to acknowledge this when dealing with interactions between people. We strive to put everyone into a very few categories.

It is a way of understanding and accepting the fact that we cannot expect to find that others behave exactly the way we do or to agree with our choices of what is good and right. This, of course, means we must each modify our choices to achieve greater agreement, and less conflict in our interactions. What choices and rules apply when our positions overlap? This is the challenge the world offers us.

4. Biblical "contradictions"

Remember those examples in Ch. 1 of "contradictions" in the Bible? We can now see an answer to some of them. God is telling us there is no single solution applicable to every variation of a given situation.

B. Conscious

"Awareness" has more to do with sense perceptions: I am aware that I am hungry. So it may be thought of as dealing with the world of the body. It may also suggest being in possession of certain information. Here it overlaps "cognizant", which relates to reason and knowledge. For example, I am cognizant of when I ate last, and realize I can deal with the hunger by eating. Now we are in the world of the mind.

"Conscious", however, means being aware at still another level of realizing some truth. One example is realizing that no matter how often I eat I will get hungry again. In another sense, it is a perception of one's Self as a distinct entity. "Conscious" also involves understanding what our civilization rests upon.

"Committed" is worth considering. Committed to our families, our neighbors, our Earth, and whatever set of values we find it possible to believe in.

We might also include the notion of "connected" as an essential part of our world view. We cannot expect to be connected each to everyone, but being connected to only one is a dyadic model of society. Some must be connected to at least two other persons, which gives us a string or network model of society. It could follow from "pragmatic".

Perhaps "constructive", "consistent", "considerate", or "compassionate" are desirable attributes. It is a question which characteristic deserves the greater emphasis, as being more fundamental. For now, I am inclined to think these follow from being conscious.

C. Total

Eclectic Pragmatism is intended to include everyone, be applicable to everyone. Otherwise, we will have only another set of tribal rules. It is also intended to consider all ideas (experiential, philosophical, and spiritual) in the process of selecting those that help us deal with our problems. This does not mean we accept into practice all ideas and notions. It means we are willing to examine them, to foster those of greatest promise, and to set aside those that conflict or contribute less to achieving our goals.

> **Comment:** There is a serious human problem here. It seems for any given situation, there will always be some arguing adamantly on opposite sides of the question. History has shown that arbitrary use of majority rule is not an infallible basis for deciding such differences. In such cases, a second level rule might be to minimize harm to the minority view. However, a society clearly cannot let a minority viewpoint always prevent action.

"Trusting" merits consideration. It may be a dividend from practicing all the other characteristics. Nevertheless, some may have to risk a loss by starting.

D. Responsible

This means accepting authorship for the consequences of one's decisions and actions. Without it, the ripples we make come back to us as tidal waves of indignation, anger, and violence.

"Accountable" is said to be synonymous with "responsible". I prefer "responsible" because to me it means accepting the consequences of one's actions regardless of whether there is anyone to be accountable to. We may want to consider "reasonable".

Charles H. Peterson

E.   Enterprising

This signifies an active approach to our relationships with each other and with our Universe. It means reaching out to deal with situations rather than letting them fester and eventually overwhelming us.

Other qualities could have been chosen, such as "caring", "persistent", or "related". The ones chosen fit the acronym

RESPECT,

which means to look again, at our fellow humans. "Respect" of persons involves admiration, approval, and deference. We can also respect the components of our system of beliefs. We choose to use these qualities as guides in our lives because we admire the qualities they represent, we approve of them, and we defer to them. "Defer" means we yield to them as opposed to insisting on gratifying our own wants, cravings, and lusts. RESPECT thus covers "related" and "caring". It also covers respect for oneself.

Chapter 17. Conclusions

Science can describe and utilize natural processes, and even modify some of them. It can often relate cause and effect. A crucial limitation Science recognizes in its creeds is the principle that no idea can be accepted unless validated by measurements. Otherwise, each idea is only another speculation and acting on it a matter of faith. What we have found is:

• Science has many instances of faith in its creeds.

• Science has failed to explain the ultimate nature of matter or the Universe or to explain our origins.

• If our Universe began in a Big Bang, no one has been able to conceive a way of making measurements of the conditions before our Universe began.

Science thus is illogical. It proudly points to all the regularities ("Laws") it has uncovered (not invented) in the material portion of our Universe. We say a proof in mathematics requires both necessary and sufficient conditions. These "Laws" may be only the necessary conditions for an explanation of our Universe. But they are not sufficient for a complete explanation. They are empirical approximations with arbitrary constants of proportionality.

Science shuns investigations in the spiritual portion, arguing that there is nothing to measure. It would appear more reasonable and logical to assume that that sphere would also show regularities (Laws).

Secularists say Science proves there are no absolutes, or at least has found none. We have seen this is absolutely incorrect: atomic structure, crystal structure, Avogadro's Number, the charge on an electron, the Law of Gravity. So if the material world has absolutes, why should we not expect to find absolutes in the world of the mind and in the world of the spirit?

Question: Science asserts there is only the material world. Then why does it insist its pronouncements in that world are applicable to the mental or the spiritual world? And even in its proper sphere, its versions of Natural Laws are often limited to portions of the ranges of the variables.

Science is also asking us to believe ever more unreasonable things, while its sages wag their heads in unison: quarks, gravitons, strings, monopoles, curled up undetectible dimensions.

It is argued that we are all different. In Ch. 16 we noted why we have reason to believe we are unique individuals. I suggest this argument is overdone. We need to recognize a middle course between being unique and being a clone. We already do this by acknowledging similarities we consider important.

In another regard, if we are all different, why do so many lie, steal, lust, rape, and murder? This surely is replicated behavior.

One of the pictures that comes to mind is that of our life process. We ingest certain substances we call food from our environment. Our bodies extract nourishment and energy from these foods. We excrete waste to the environment. What is the meaning? We are merely waste makers. If we do nothing more than this, we are only one step away from being the sewer itself.

One conclusion is absolutely certain: if the Universe we can measure is all there is, we will get <u>absolutely</u> no help from it. Such a Universe is <u>absolutely</u> indifferent to whether we live or die. For what we are doing to our Earth, which is our home, the Universe is probably better off without us.

Next, Science displays most of the important characteristics of a religion in having sets of:

- Dogmas (Big Bang, Evolution, Relativity, exchange particles, quarks)

- Doctrines (Uniformitarianism, Expanding Universe, Random events, Spin, the Lorentz Contraction)

- Taboos (God, supernaturalism, time prior to the BB, miracles, psychic phenomena)

- Priests (experts, interpreters, spokespeople)

- Values (rationality, consistency, materialism)

- Rituals (conventions, technical meetings, honors)

We can list some of the beliefs as follows.

A proton	baii[1]	is a point charge.
		is a point mass.
		is made of quarks.
An electron	baii	is a point charge.
		is a point mass
		is a wave.
A photon	baii	has mass.
		does not have mass.
Light	baii	is a wave.
		is made of photons.
Gravity	baii	is a space warp.
Space	baii	is expanding.

[1]baii = behaves as if it

We have already noticed that both Science and Religion agree that the first event in the development or creation of our Universe was the appearance of light. As I read over what I have written, another similarity somehow came into my mind. Consider these references:

- Tabernacle

  Exodus 25-27: The LORD commanded Moses to build a tabernacle to contain an ark into which was to be placed the Tables of the Law.

- Rock

  Deuteronomy 32: 4 "He is the Rock. . . :

  Deuteronomy 32:31 ". . .their rock is not as our Rock. . . ."

  Comment: Meaning, the pagans worship powerless stone gods.

Remember the standard of mass adopted by Science? That arbitrary chunk of platinum-iridium guarded so carefully in some tabernacle (building) near Paris? It is the "rock" on which the pagan Religion of Science rests.

A third conclusion: Science has absolutely <u>nothing</u> to offer us in our human relations. The most obvious reason this is true is that Science insists on measurements as the only admissible proof of an hypothesis.

So far no one has found a way to measure the nonmaterial elements in our Universe. Science thus declares itself incompetent to analyze and evaluate these nonmaterial worlds. It has made itself irrelevant to the task of resolving human problems. And we have all been misled and cheated by its claims.

We have passed the fork in the road. Some of us are becoming aware that we are on the wrong path. We have made a Type A error: the answers we seek are not to be found in Science. It can only give us answers based on what it can measure. Science itself confesses there are many things it cannot measure.

Let me suggest some perspectives.

1. The collapse of Communism in Soviet Russia after 70 years of brutal experimentation with their own people left millions with a hopeless, hollow hunger for something to believe in, something around which to organize one's life, that is again finding expression in Religion.

    Comment: Which is why I shuddered when I probed into Dewey's experimentalism.

2. Perhaps the thought that there must be something to believe in shows there is a bit of God in all of us, or at least something that can respond to the idea of a God.

3. It is not that it is for us unthinkable that there is no God. Rather, it is unthinkable that there could not be a God.

So my main conclusion is:

It is not a question of whether there is or is not a God. It is rather a question of what kind of God do we want to believe in and follow.

But that is another subject. . . .

Meanwhile, let me add a few more characteristics of EP. Some of the characteristics pertain to EP itself while others apply to the person practicing it.

A. Voluntary

History has shown numerous organized religions. I may be wrong and harsh in this judgment, but it seems to me most, if not all, have been corrupted by humans into unforgiving, oppressive systems. The Bible, however, has pointed out the central role of <u>individual</u> faith. Acceptance of a set of beliefs is, I believe, and <u>must</u> be an individual decision, an individual act of will.

This does not mean that each of us is free to put off the acceptance indefinitely. We need principles of behavior. Not man-made laws. Man-recognized Laws. We must each actively examine the alternatives as shown by human experience, which gives us the raw data as to our nature. Where that nature comes from is another question. But we do have one.

And we must consciously choose what we will believe in. We must have some set of beliefs, otherwise we will flutter like a windblown leaf from path to path. We could have no lasting, mutually uplifting relationships for no one could rely on our behavior from one minute to the next. We should accept the basic principles well before becoming adult. As adults we can continue to increase our understanding of them.

B. Interdependent

"Connected" is insufficient. We must use our ability to be conscious that we may realize how much we actually depend on others and our environment.

C. Accepting

While it is commendable to strive, there are times when we need to limit our rebellion against circumstances. We have no way of preventing large scale changes in our environment, like drought, flood, earthquake, etc. These lead to changes in our social organization. We must be willing to change in the face of change. The alternative is to perish. Thus we must be adaptable.

We must then accept the most promising course of action available to us and not grieve over unavailable alternatives. Since "adaptable" could be covered under "eclectic", I favor "accepting" as a principle. We discussed "accountable" under "responsible".

D. Giving

We can give and we can take. If we only take, the result will be a drying up of giving. The oceans "give" us rain, which we can use for drinking water and as a source of power in hydroelectric plants, and even still as a source of food. But eventually we "give" it back to the oceans. Fortunately, we cannot prevent this.

Charles H. Peterson

E. Open

"Open" includes "open-minded" but also open emotionally and spiritually. Some might think characteristics such as "organized", "observant", and "objective" are more important.

F. Directed

"Directed" means having a sense of progressing through life toward some goal. It implies "determined", a commitment to helping achieve that goal. "Determined" also seems more active compared to a passive "devoted". "Disciplined" is perhaps another desirable attribute, but it seems covered by "Responsible".

These expand the acronym describing EP to:

RESPECT VIA GOD

REFERENCES

ACEA A Concise Encyclopedia of Astronomy. A. Weigert and H. Zimmermann. American Elsevier Publishing Company, Inc. New York. 1968.

AEU At the Edge of the Universe. Alan Wright and Hillary Wright. Ellis Horwood Limited. Chichester, England. Simon & Schuster International Group. 1989.

AIPH American Institute of Physics Handbook. Second edition. McGraw-Hill Book Company, Inc. New York. 1963.

AJASC Atom: Journey Across the Subatomic Cosmos. Isaac Asimov. Truman Talley Books/Plume. Penguin Books. New York. 1992.

AME Advanced Mathematics for Engineers. H.W. Reddick (New York University) and F.H. Miller (The Cooper Union). Second edition. John Wiley & Sons, Inc. New York. 1947.

ASTM Standard for Metric Practice. E 380-76. American Society for Testing and Materials. Philadelphia. 1977.

BDS The Birth and Death of the Sun. George Gamow. A Mentor Book. The New American Library of World Literature, Inc. New York. 1952.

BHQU Black Holes, Quasars, and the Universe. Harry L. Shipman. Second edition. University of Delaware. Houghton Mifflin Company. Boston. 1980.

BHT A Brief History of Time. Stephen W. Hawking. Bantam Books. New York. 1988.

BTGP Basic Teachings of the Great Philosophers. S.E. Frost, Jr. Barnes & Noble, Inc. New York. 1942.

CAE Creation and Evolution. Alan Hayward. Bethany House Publishers. Minneapolis. 1995.

CEH Chemical Engineers' Handbook. Fifth edition. Robert H. Perry and Cecil H. Chilton, editorial directors. McGraw-Hill Book Company. New York. 1973.

CP Conceptual Physics. Sixth edition. Paul G. Hewitt. City College of San Francisco. Harper Collins. 1989.

CPP Chemical Process Principles. Thermodynamics. Ch. XI. Olaf A. Hougen and Kenneth M. Watson. John Wiley & Sons, Inc. New York. 1947.

CPP Chemical Process Principles. Thermodynamics. Ch. XVII. Olaf A. Hougen and Kenneth M. Watson. John Wiley & Sons, Inc. New York. 1947.

CRC CRC Handbook of Chemistry and Physics. 6th edition. 1984-1985. Robert C. West, editor-in-chief. CRC Press, Inc. Boca Raton, FL.

CREA Creation: The Story of the Origin and Evolution of the Universe. Barry Parker. Plenum Press. New York. 1988.

CVEP A Contemporary View of Elementary Physics. Sidney Borowitz & Lawrence A. Bornstein. New York University. McGraw-Hill Book Company. New York. 1968.

DE Darwin's Enigma: Fossils and Other Problems. Fourth edition. Luther D. Sutherland. Master Book Publishers. Santee, CA. 1988.

EAM Energy and Matter: Building Blocks of the Universe. Charles B. Bazzoni. University of Pennsylvania. The University Society, Inc. New York. 1937.

EST Encyclopedia of Science and Technology, 7th edition. McGraw-Hill Publishing Company. New York. 1992.

EU Exploration of the Universe. Second edition. George Abell. University of California. Holt, Rinehart and Winston. New York. 1969.

FG The Fingerprint of God. Hugh Ross. Promise Publishing Company. Orange, CA. 1991.

FQ Familiar Quotations. John Bartlett. Fourteenth edition. E.M. Beck, editor. Little, Brown and Company. Boston. 1968.

GNP God & the New Physics. Paul Davies. A Touchstone Book. Simon & Schuster. New York. 1983.

HEF Handbook of Engineering Fundamentals. 1st edition. O. W. Eshbach. John Wiley & Sons, Inc. New York. 1936.

HET A History of Educational Thought. 3rd edition. Frederick Mayer. The Charles E. Merrill Publishing Company. Bell and Howell Company. Columbus, OH. 1973.

HHTF Handbook of Heat Transfer Fundamentals. 2nd edition. Warren M. Rohsenow, James P. Hartnett, and Ejup N. Gani, editors. McGraw-Hill Book Company. New York. 1985.

HWP A History of Western Philosophy. Bertrand Russell. Simon and Schuster. New York. 1945.

IMW Intelligence in the Modern World. John Dewey's Philosophy. Joseph Ratner, editor. The Modern Library. Random House, Inc. New York. 1919.

IOYO I'm OK, You're OK. Thomas A. Harris. Harper & Row. New York. 1969.

KJV The Holy Bible. King James Version. Thomas Nelson Publishers. Nashville. 1976.

KTG Kinetic Theory of Gases. Earle H. Kennard. 1st edition. McGraw-Hill Book Company, Inc. New York. 1938.

MHG Man and His Gods. Homer W. Smith. Grosset & Dunlap. New York. 1952.

NCE Nuclear Chemical Engineering. 2nd edition. Manson Benedict (Massachusetts Institute of Technology), Thomas H. Pigford (University of California, Berkeley), and Hans Wolfgang Levi (Hahn-Meitner-Institute für Kernforschung Berlin). McGraw-Hill Book Company. New York. 1981.

NIV The NIV Study Bible. New International Version. Kenneth Barker, General editor. Zondervan Publishing House. Grand Rapids, MI. 1985.

PCAI Principles of Anatomy of Invertebrates. W.N. Beklemishev. The University of Chicago Press. Chicago. 1969.

PHT Process Heat Transfer. Donald Q. Kern. McGraw-Hill Book Company, Inc. New York. 1959.

PITC Philosophy in the Twentieth Century. Volume One. William Barrett (New York University) and Henry D. Aiken (Harvard University). Random House. New York. 1962.

PSY Psychocybernetics. Maxwell Maltz. Pocket Books. Simon & Schuster. New York. 1971.

QCIS Quality Control and Industrial Statistics. Acheson J. Duncan. The Johns Hopkins University. Revised edition. Richard D. Irwin, Inc. Homewood, IL. 1959.

RHCD The Random House College Dictionary. Revised edition. Jess Stein, editor-in-chief. Random House. New York. 1980.

RSGT Relativity: the Special and the General Theory. Albert Einstein. Crown Publishers, Inc. New York. 1961.

SE Situation Ethics. The New Morality. Joseph Fletcher. The Westminster Press. Philadelphia. 1966.

SHH Science Held Hostage: What's Wrong with Creationism AND Evolutionism. Howard J. Van Till, Davis A. Young, Clarence Menninga. InterVarsity Press. Downers Grove, IL. 1988.

SSE Solar System Evolution. Stuart Ross Taylor. Cambridge University Press. England. 1992.

STR Space, Time and Relativity. H. Horton Sheldon. New York University. The University Society, Inc. New York. 1935.

TCOE The Collapse of Evolution. 2nd edition. Scott M. Huse. Baker Books. Grand Rapids, MI. 1993.

TDU The Dynamic Universe. 3rd edition. Theodore P. Snow. University of Colorado. West Publishing Company. New York. 1988.

TEU The Early Universe. E. W. Kolb and M. S. Turner. Addison-Wesley Publishing Company. Reading, MA. 1993.

TFE The Fifth Essence: The Search for Dark Matter in the Universe. Lawrence M. Krauss. Basic Books. Harper Collins. 1989.

TGB The Golden Bough. Sir James Frazer. Wordsworth Editions Limited. Cumberland House. Hertfordshire, England. 1993. First published 1890.

TGP A Treasury of Great Poems. Louis Untermeyer. Simon and Schuster. New York. 1942.

TGR The Global 2000 Report to the President. The Technical Report. Volume Two. The Council on Environmental Quality and the Department of State. Gerald O. Barney, Study Director. 1978.

TMEW The Meeting of East and West. F.S.C. Northrop. The Macmillan Company. New York. 1947.

TNU The Nature of the Universe. Fred Hoyle. Cambridge. Harper & Brothers. New York. 1950.

TOG A Text-book of Geology. Louis V. Pirsson and Charles Schuchert. John Wiley & Sons, Inc. New York. 1915.

TOP The Omega Point. John Gribbin. Bantam Books. New York. 1988.

TPCE The Principles of Chemical Equilibrium. 4th edition. Kenneth Denbigh. Cambridge University Press. Cambridge, England. 1981.

TR The Red-Shift. Allan R. Sandage. The Scientific American. September 1956, pp 171-182. Scientific American, Inc. New York. 1956.

TSC The Story of Civilization. Vol. II. The Life of Greece. Will Durant. Simon and Schuster. New York. 1939.

TWGT The World's Great Thinkers. Man and the Universe: The Philosophers of Science. Edited by Saxe Commins & Robert N. Linscott. Random House. New York. 1947.

TWM Goedel's Proof. E. Nagel and J. R. Newman. The World of Mathematics, Volume III. J.R. Newman. Simon and Schuster. New York. 1956.

UP University Physics. Francis Weston Sears, Massachusetts Institute of Technology, and Mark W. Zemansky, College of the City of New York. Addison-Wesley Press, Inc. Cambridge, MA. 1950.

WTID Webster's Third New International Dictionary of the English Language Unabridged. Editor-in-chief Philip Babcock Gobe. G & C Merriam Company. Springfield, MA. 1966.

ECLECTIC PRAGMATISM

Index

Charles H. Peterson

About the Author

Mr. Peterson (MChE, MIE, MPIA) draws upon an eclectic analytic background as engineer, stockbroker, soldier, medical technician, questioning Christian, and parent to highlight inconsistencies and unanswered questions in Religion and Science. He has heard, as Omar Khayyám wrote, "great argument about it and about, but evermore went out by the same door wherein I went". Many descriptions of the problems of society exist, but little in the way of widely accepted solutions. Several dictators have promised solutions based in part on Darwinian assumptions only to succeed in just a few decades in killing off well over 100,000,000 of their own people before they themselves passed from the scene. If you value your freedoms, it is still not too late to join in the effort to understand the bases for those freedoms. As a start, his *Eclectic Pragmatism* aims to identify what we can accept from the assumptions behind the explanations offered by Science and Religion for our Universe.

www.ingramcontent.com/pod-product-compliance
Lightning Source LLC
Chambersburg PA
CBHW031815170526
45157CB00001B/66